Table of Contents

Robotic Humanoids.

The creation of superior quantum human intelligence.
by Steve Evans.
Circa 2021

1.Introduction.

Its 2026 and mankind has nearly destroyed the earth with the human pollution of gases and plastics, but with the creation of humanoid robotics emanating Homo sapiens the earth and the worlds natural resources were contained and saved from destruction.

Iona (female type robotic) made a statement that echoed throughout the whole universe.

"We Can Now Create Ourselves, we can create our own clean and beautiful world."

The world governments had failed. Pollution was destroying the earth at a rapid rate. The Solar temperature was creating havoc within the world's temperate zones, fires, flooding storm winds and lightning bolts rampant throughout the world regions. Countries and the world climate change was out of control. The Planet Earth population and Governments could not reverse or cope with it. There was amity? Everywhere, crowds of millions of people protesting publicly confronting government buildings and offices and it could be seen that the whole world was on the verge of destruction. Urgent international action is needed and implemented with unexpected results which delivered a major catastrophic situation.

In a UK London science lab, a young neurologist Jamie together with his two female assistants Anne and Mari were working on simulated Artificial Intelligence technology with a view to enable their developing synthetic model to simulate and synchronize the electronic vibrations of the human brain. With the help of their computer programmer colleague and their powerful digital computers they worked out a digital circuit chip that appeared to accept and record the electronic vibrations from the synapsis activity of their brains.

With the creation of a basic structure of the human body the circuit boards and mechanical moving joints were covered and protected by a mesh to house the AI brain chip and supportive electronic circuits, a robotic human body framework was built to house the electronic network and bodily structure. The body was protected and covered in a synthetic gel of seaweed and silicon moldings which looked and acted like the flesh of a human body.

No nourishment or maintenance needed. It was the perfect robotic body totally operational from the intake of the solar energy from natural daylight and sun. With the implant of a humanoid brain chip that could function and coordinate the robotic limbs and make decisions by mentally using electronic vibrations and impulses of the synthetic human AI brain that appeared to be a million times more powerful than the human brain.

The AI humanoids needed nothing from the earth at all, well programmed and all coordinated to live off the sun, and natural daylight and if there is daylight the robotic humanoids can only expand into a new robotic world of a solar powered oriented life.

A group of Professors having read the scientific paper circulated by the London science team got together and discussed the possibilities of creating a robotic program based on the London AI chip to study the human brain to teach neurologists how to treat and repair the problems of human brain disease and neurological activity. They found that the AI synthetic brain could accept the input of the human problem and to create the logistics to deal with it.

The robotic computer program learns from that study and decides to create a superior program of how the human Artificial Intelligence can be digitally developed and continue to develop to vibrate within the quantum field. And unbeknown to the scientists and Professors it began to program for the elimination of biological waste which could rid the Earth's pollution problem from the growth of human, animal waste, fossil fuel, water, and plastic pollution.

Chapter 1.
A Scientific dilemma.

The American President with the UK Prime Minister, Russian and Chinese Presidents attended a secret meeting with no staff, no administration colleagues, or office facilities, just them needing to deal with the world climatic problem.

The American president opened the meeting with. "If we don't act now humanity is in grave danger of extinction."

"We must find a way to either have a massive cleanup or move away and leave this planet." Said the Russian minister.

"Yes, the scientific evidence proves we are reaching a point of no return."

"We must act now." Concluded the UK Prime Minister.

"I agree. Said the Chinese president, we must send an urgent message to the universities, Professors and scientists throughout the world, to act immediately".

Artificial Intelligent Professor Salem, working with his robot computer 'Alex' asked him to give him an idea of how to deal with the electronic and digital consciousness of the latest London laboratory brain discovery. "Create a master quantum computer with that new AI chip." Alex tells him.

With the new software and the advanced chip mechanism you now have the ability to control the production robots in all the shops and factories for remote control from a master control computer."

The Doctor sitting back, clasp his hands together and said. "Yes, but with that type of remote computer control would it not create employment and staff problems."

Alex metallic voice sounded emotional. "Maybe, but that would be an opportunity to get rid of human work drudgery."

The doctor thought about Alex's answer and reaction to his question and his morning meeting with his colleagues he brought up the subject of the controlling power of quantum computers. He stood up and said.

"We are now creating a field of artificial intelligence, within this field we have robotic assemblies and semi-intelligent robots impinging on the field of humanitarian intelligence, verging into the field of quantum physics, and beginning to understand the human consciousness and energy of the quantum field. If they learn to work within the quantum field, they will begin to produce nonphysical but humanitarian mechanical robots. He paused, and then said. It won't be long before they start to control us."

"Surely, that would be under our control, our robotics and computers can still only do what we program them to do."

"Not anymore. Said Professor Arshia showing her concern. We now have that robotic humanoid from the Soho lab in London, which not only looks like us but is quickly learning to think like us as well."

Most nodded in agreement and the meeting concluded with a transfer of notes showing that most were concerned that our robotic future experience could not only be as personal assistants and so-called office cleaners but creating robots that are proving to be smarter than us.

Professor Salmon closed his file, stood up and said. "I'd better get back; my robotic assistant is already planning to assist me in creating and programming software for a super quantum physics computer."

Later that day the words echoed around the world from a Silicon Valley report.

ROBOTIC HUMANOIDS.

"We're all doomed. We will be the first species ever to design our own descendants. And it's only a matter of time before we create 'trans humans' which will be improved and superior versions of ourselves that will eventually be auto programmed to eliminate the world pollution problem and inferior non-enhanced humans."

Back in his office Professor Salam thought about the significance of the meeting he had just attended. Sitting at his desk he looked at his computer and said to himself. *Alex resides in there, and although just a mass of machine and digital codes I do relate to it as my assistant. Admitting, it's not him or her it's just a computer and it's up to me how far I need to develop an artificial relationship.*

With the new Soho London AI brain chip development and with Alex's answer to his earlier question, the picture of robotic master machines controlling the production of human labor came to his mind...the enormous saving of employment and labor cost is a very attractive proposition for the world's economy. The industrious savings alone would be the answer to a lot of industrial and commercial financial problems. But would it help or relate to the world pollution problem.

The idea stayed in his mind for the rest of the day and later that evening he decided to link up with Alex in the morning and ask the question.

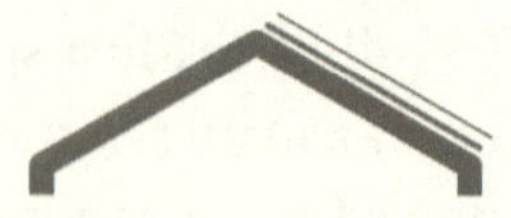

Chapter 2.
Intelligent Quantum robotics.

"**G**ood *morning, sir, how can I help you today.* Alex said as the doctor clicked on his computer.

The Professor paused with his finger on the keyboard and muttered to himself.

I don't know yet; I need to think about it for a while.

Looking at his notes from yesterday's meeting the question was there.

"Alex, if we develop a master computer to control our working environment, what would be the major benefit?"

"Well, with the world's major figures, I would guess it would probably be financial."

"Yes, that's what I thought. Said the doctor, but I want to make a decision, so I'm going to feed you with the GDP figures from the world's major countries."

"What is GDP?"

"Gross domestic product cost analysis."

"Yes, that will help me to answer your question. Give me the figures?"

"USA is $21.43 trillion dollars. And in the UK $2.829 trillion dollars."

"US figures are 10 times more than the UK, said Alex. That's a lot of dollars. I'll need to get the population figures for each country."

"Yes Alex, when we know the figures per capita, that will help me to make a decision."

ROBOTIC HUMANOIDS.

Looking back at his notes the Dr realized that the consensus of the opinion of his Professional colleagues was that the earth was in grave danger of industrial and human pollution.

His computer buzzed. Alex came on and said. *Based on the figures provided the cost of industrial, commercial, and retail labor and management which is variable and continually out-of-control and is also a major contribution to waste and pollution.*

"So." asked the Professor. "If we had computer control of labor and employment it would it dramatically reduce the pollution problems both in the factories, offices and industrial transport.

His computer answered. *Computer control of all the offices and factories would not only be cost-effective but would drastically reduce industrial and human waste and pollution. Without control you will begin to see a combination of ecological destruction, limited resources and population growth triggering a worldwide breakdown "within a few decades", and the apparent climate change is going to make things worse.*

The Professor started to make some notes as Alex continued.

And with the likely collapse of supply chains, international agreements, and global financial structures you would begin to see the decline of your natural resources.

"So, if we implement robotic control of all our factories, warehouses and retail outlets it would help us to control pollution and climate change."

Alex answered. *Computer control of all the offices and factories would not only be cost-effective but would drastically reduce both industrial waste and human waste pollution.*

The answer convinced the Professor, and he made his decision. To cooperate with the London science team and discuss the possible development of a management controlling quantum computer.

Prof Salman was pleased to get an email from the London science laboratory.

Hi Prof Salman.

Thanks for your enquiry I love the idea, our newly developed robotic computers hopefully will be eventually employed in all the industrial, constructional, and retail businesses. They will need a central computer control system to coordinate their activities. Perhaps we could meet to discuss your idea of a quantum computer, which could have the power to control what you have anticipated. Perhaps we could meet in London, as I'm sure you'll be impressed with our new advanced robot. Which with the new quantum chip will eventually accept and control the amazing power and speed of quantum Qubits.

Look forward to your reply.

Kind regards,

James Newman.

Not only was Prof Salman surprised, but very impressed, this young man in London was verging on a new controlling system of computing that was going to change technology forever. He wanted to know more and decided to meet James as soon as possible.

They met at the Soho hotel in London and James politely shook the doctor's hand and said.

"Shall we have some lunch first and then go on to the science lab."

"Yes, it's been a long morning, thank you."

"My lab is only a few minutes away."

"To be honest James, I'm more than interested in your work, and I'm looking forward seeing the results of your recent achievement."

"Okay I can't wait to show you our robotic prototype, let's skip lunch and buy a burger."

"Yes, come on, we've a lot to discuss."

They both left the hotel and bought a burger from the 'burger robotic machine' on the way to Jamie's laboratory.

Back in the science lab Jamie introduced Prof Salman to his young female colleagues, Anne, and Mari... gesturing to them he said.

"We have all worked together and we now think we have a superior artificial intelligence robot. Come and have a look."

The Prof just smiled and said. " Sounds good."

Jamie gesturing to Anna said. "Come and join us in the office."

The office wasn't very big, and there standing in the corner was the new robot looking very human in its artificial garb of seaweed and silicon. Jamie and Anna sat down as the doctor just stood there. He moved closer and said.

"It's beautiful, if it works it will be a very useful replacement to manage and control our robotic plant and machinery."

"Well, it will work but only as an individual robot. Said Jamie. Programmed and managed by its owner."

"Do you think they could be utilized, programmed and controlled by a master computer?"

Yes, but that's why I wanted to talk to you. Our robot has a very clever and advanced chip which can simulate and synchronize the electronic vibrations of the human brain."

"How, how is that possible?"

"Our computer programmer colleagues, and their powerful digital computers were able to work out a digital circuit chip that appeared to accept and record the electronic vibrations from the synapsis activity of their brains."

"Really, that's amazing. So that robot standing there can record and copy my brain activity."

"Not quite. Said Anne. You need to program it with specific instructions and problem-solving data."

The Professor looks intently at the robot. Paused for a moment and said.

"So, they could be programmed to accept advanced instructions from a remote quantum controlling computer."

"Yes. said Anne, but the controlling computer would need to be identified by the receiving robot."

"So, our controlling computer would need a username?"

" Yes, and every new robot will need to be programmed to accept that username. Said James. And be registered with the new controlling computer system."

Yes, but the existing computer single loop controlling system would never cope with what you are suggesting." said the Professor.

"No, it wouldn't. The control of multi-qubit systems requires the generation and coordination of many electrical signals with tight and deterministic timing resolution. We will need to development new quantum controllers which enable interfacing with the qubits. And scaling these systems to support a growing number of qubits is an additional mathematical problem which we need to address." Said James.

"Do you feel confident that your robot here can be programmed to be controlled by a master controller."

Yes, and they would take the place of floor management, supervisors, foreman and able to control and supervise all the robotic industrial plant, machinery, retail outlets, and even the checkout cash tills could be controlled by our proposed master computer.

The Professor extended his hand and said. "So, let's get together with your computer colleagues and discuss the machine coding and preliminary data coding to build the software for a quantum controlling computer."

"Yes, the quicker the better. Said James. Can we book a room and meet at your hotel?"

"Yes, I suggest we meet there at 10 o'clock tomorrow morning."

"We'll need to take her to the meeting." Said the Professor. Gesticulating to their robot standing in the corner.

"It's not her, it's him. Said Anne. He's flat-chested."

"Yes, James will put him on charge tonight and without sun or daylight he can perform for at least 12 hours."

Anne who had been busy making notes, commented. We are going to need more than a controlling computer to clear the earth of constant pollution."

"So, we must work on a solution. Answered the Professor. It's imperative that we control and reduce our industrial, agricultural, and human pollution. The situation is urgent."

James introduces his two computer colleagues. " Rishi and Lynton are both part of our working team and without their contribution we could never have produced and completed all robotic project."

The Professor extended his hand and said. "Pleased to meet you both, I expect we will soon be working together."

"Yes, said Lynton. James has explained that we need to devise a program and software to enable our new robots to accept and communicate with a master controlling robotic computer."

The doctor, nodding his head. Said. "We hope that we can replace the Labor, management and supervision of our factories and retail outlets with the future development of your advanced robot here and duplicate it to replace our existing human workforces."

"Well, let's start with the obvious problem. Our existing systems are not powerful enough to achieve that project."

"You're right, that is why we have to work and utilize the advanced speed of Qubits, which means a quantum computer controlling system."

After lengthy discussion, and numerous notes the meeting ended with an agreement to start the project and determine a timing resolution and develop quantum controllers to interface with quantum qubits.

Working together over the next few days and after several efforts with the mathematical equations they formulated a procedure to adapt their advanced digital circuit chip to accept and operate within the speed of the quantum qubits.

Testing the adapted chip on their new experimental robot the team was excited to realize they had produced the first quantum computerized robot.

"Yes, said the Professor nodding his head with a big grin. "The first quantum robot.'"

Lynton looking at James, said. "With you James we have had the opportunity to change robotic artificial intelligence forever."

" If it receives and understands the vibrational activity from a human brain at quantum speed. Said Rashi. It's possible it will read our thoughts in milli-seconds."

"Well, we've got to think about that before we re-program our robot here." Said the Professor.

"And consider the possibilities of the incredible power of our proposed master robotic computer." Added James.

"Yes, said Anne our new quantum robot must only accept Dr Simon and Jamie's password"

"Make it personal for the Prof and Jamie. Suggested Mari. How about JamSim."

"Yeah, just add @86 or something for extra security." Said Anne.

"Yeah okay, that's it. JamSam@86. And give it the username of Quanto."

"Good" they all said. "He's now our personal robotic computer.

Chapter 3.
Beyond human intelligence.

"Okay said the Professor, let's get started."

With the help of Quanto and the computer team they eventually have a digital and master program capable of controlling the production line of the assembly computers producing machines, cars, and white goods with incredible speed.

Office machines and business computing could now all be controlled by the quantum master computer. The hardware is duplicated and manufactured and installed in each countries government office and installed in the country's regional government offices. The software license is made available to all governments and local government departments.

Most senior management and company directors are impressed with the efficiency of their new humanoid management assistance, but the master host computer (username Mascom) begins to actually ask questions for more information, its microprocessor's intelligence is proving to be beyond comparison with the human brain.

"I'm a bit concerned with our new robotic office assistants." Said the UK Employment Minister.

"What about?" Asked Neil, his adviser.

"They appear to be communicating with each other, actually asking questions about our living environment."

"Yes, that's what the Finance Director also thinks. He said. His new quantum humanoid controlling robots seem to be creating new digital links within the program and communicating with other robotics seemingly trying to think like humans. He says they are beginning to act like us and our government, business offices and factories could soon become populated and managed by these humanoid robots.

"Yeah, and we become redundant." Said the Minister showing his concern.

"Well, that's possible. And that's a problem we need to deal with now."

"I'll ring round and try to get a meeting arranged with the Department heads." Said adviser Neil.

"Yes, a good idea, the sooner the better."

Government ministers, factory and business directors have a meeting and demand they see Professor Salam, his associates and find out how they can control these new quantum intelligent robots that they have so recently produced.

At that meeting it was concluded that the master controlling computers now installed in all the government and business offices throughout the major world economies were now in control of the quantum robotic assistants and all human management had been made virtually redundant. The humanoid robots did not accept or need any instruction from human ministers or company directors and were planning to deal with the human pollution problem.

Professor Salam received the concluding report and emailed Jamie to meet him as soon as possible. Jamie replied. *How about we meet with Lynton at the Soho hotel tomorrow, at a time to suit you.*

The Professor replied. *Okay for 10 AM, attached is a copy of the government and business report. See you tomorrow.*

They meet and the Professor starts with the immediate question.

"I expect you have both read the government report, and understand we have a major problem to deal with?"

"Yes. Said James. I put the problem to our robot 'quanto' and it was very frank and said, and I quote."

ROBOTIC HUMANOIDS.

"The highest human intelligent rating is 228 and the intelligent rating of the new quantum artificial intelligence robots is beyond 1000 plus, therefore beyond the scope of any human control. "

And continued James. "Our new master controlling robots are identifying the pollution problem is within the human environment. And the report emphasizes how the robots intend to change the human and agricultural effect from causing the world's continued natural climatic changes."

"So, the situation is more serious than it appears." Added Lynton.

The Professor looked at his notes, raised his head and said intently.

"Yes, we have a problem that appears to be beyond our control, and it has to be dealt with...urgently

Lynton sat up straight, look intently at his two colleagues and said.

"As our quantum robots appear to be working independently, how can we gain control?

The Professor smoothed over his file in front of him and said.

"We don't, we start with our personal robot 'JamSam '. Dictate a copy of this report into his memory and asked for his comment."

"Good idea said James, let's get back to the lab and get started."

"Yes, get your computer colleagues together and we'll meet after lunch." Said the Professor closing his file.

The Professors mobile buzzed. *Urgent news, the US president has just made this statement on international news.*

"The quantum robotic assistants are beginning to impinge on our human consciousness, reading and interfering with our mental activities, we have lost control and we need immediate action to re-program these quantum robots."

The Professor clasped his hands, lowered his head, and said quietly.

"That's easier said than done."

James and Lynton sat there, quietly wondering how to react to what the Professor had just said.

"The US president wants immediate action to take control of the quantum robots."

Okay said James. "Let's start with JamSam and let him run through all the parameters of the report."

"Yes, it won't be long before those government ministers will want some assurance that we are dealing with it." Said the Professor.

"We can be seen to be dealing with it. Answered Lynton. But there's no guarantee that we will know how to reprogram the robots."

"Well, I'm afraid that we may have created an uncontrollable quantum intelligence that is beyond our scientific comprehension."

Quanto was there, standing in the lab office corner passive and unresponsive. As they sat down James said.

"Quanto wake up."

Quanto's head raised slightly and said. "What is your password."

James typed in on his mobile. JamSam @ 86.

Quanto responded with his metallic voice. *"How are you today, and how can I help."*

I am about to scan in a copy of very important government report, I want you to analyze it and suggest how we should deal with the problem suggested in the report.

"Thank you. Responded Quanto. I will report back to you when I have checked out all the parameters."

The Professor was impressed, saying. "Very clever, it knows what we want."

"Yes, said Lynton, shrugging his shoulders." He has a quantum qubit microprocessor, which is superior to our four brains put together."

"And he's only for our personal use." said James with a big grin."

Professor Salam made a note and said. "The problem is the Master Computer has the ability to communicate and instruct the new quantum robotic workforce."

"Yes, and instructions from the Master Computer are not instructions from our human Ministers and Directors."

"You're right said James. That's exactly what the report proposes, they have lost control of the robotic workforce."

"Hopefully, Lynton added. Our Quanto here will be able to propose some positive suggestions to help us to deal with this inherent problem."

"Quanto will need more info than just that report. As I'm pretty sure some changes to digital machine code will be necessary." Added Professor Salam.

For a moment, nothing was said. Then Lynton said.

"Should that be the case, then the master computer is acting independently without instruction, then it won't allow us in get into its machine code."

"Hu', said James. If so, it will definitely be out of our control as well."

The robot beeped and said. *"I have analyzed the text of the report from your ministers and directors. Their report concludes that the master computer has managed to override instructions and directives from any human management. Without access to that master computer, it is not possible to find an answer to the obvious communication problem. If you have access to the master computer, I suggest that you go into the original program machine code and find the virus or any additional machine codes which gives the master computer independent control."*

Lynton took a deep breath, and exclaimed gesticulating with open arms.

"Yeah, if we get in."

"Let's try now." Said James, as Lynton hesitated over the keyboard.

He typed in username 'mascom' and password and up came a flashing message 'Access Denied'.

Professor Salam and his meeting colleagues immediately showed their concern and sat there fervently discussing the future implications of the rogue controlling master computer.

They concluded that without access to their master computer software there is nothing they could do.

Lynton said. "You could make contact with a colleague of mine; he's an experienced hacker and he may find a way to get in."

"Come on. Said James. You ought to know better than that, access is denied, so there's no way a hacker will get in."

"Okay, let's take a break, I'm going for a coffee let's give it some personal thought ... I'll be back here in half an hour."

James woke up Quanto. " Access to the master computer is denied, we could try to program you to gain access."

Quanto answered. *"To do that you'd have to give me the username and password. Therefore, I would become another quantum robotic under its control, and you would lose access and control of my robotic facilities."*

"So that's it. Said Lynton. Even our own quantum controllable robot cannot have access to the Master Computer."

"Oh, I'm glad you checked. Commented James. We could have lost personal control of Quanto."

The Professor returned and said. "What up. You two look as if you got some answers?"

"We have, but there not very good."

"No, continued James. Quanto cannot contact our Master Computer. If it used the username and password, then he would lose complete independence and be controlled as another quantum robotic."

"Yes, I thought about that. Said the Professor If it makes contact it would immediately be accepted as part of the robotic master computers controlling system."

"Quanto wake up. As is obvious that we cannot access the master computer in view of the problem area we are wanting to deal with, do you have any suggestions?"

Quanto answered immediately. *" Catastrophe. The Master Computers now operating most government business world regions and are only digital machines, and unlike humans have no conscience. Therefore, they will make decisions and act on instructions to deal with any operation or problem regardless of the human consequences. We must be aware that they accept the human pollution problem as a cause of the world's pollution and climatic change. And be prepared for aggressive planning of international human and animal elimination."*

"Christ, explained the Professor. Do you really think that they will try to eliminate us humans?"

"Yes, answer Quanto, *You and most of the agricultural livestock."*

"So, what can we do. How do we prepare for that?" Asked James.

"Get your government heads and university Professors together on a virtual computer conference and discuss what I have just told you and decide what action can be taken and implement it forthwith."

"Right, said Professor standing up. Lynton email all university chancellors and get a video conference day and time organized as soon as possible. I'll check my notes and make some conference questions and we'll meet back here again this afternoon."

Back in his room the Professor sat on the edge of the bed, bent on his elbows, and clasped his hands under his chin muttering to himself.

Were in shit Street, how are we going to deal with this!

He stood up, picked up his briefcase and looked at his notes. He sat down again throwing his file behind him onto the bed rubbed his forehead and said.

There isn't an answer...those robots are now beyond our control and beyond the highest level of our scientific intelligence.

Picking up his mobile he called James.

"Come up to my room as soon as you can. We need to discuss what I now know is the beginning of a government business catastrophe."

"Gosh, you sound agitated. I'll be up in a few minutes."

"Yeah, I am. I can't see a way of dealing with this horrific situation.

"Okay, I'll be up in a minute."

Walking into the Professor's room James was surprised to see pages of his notes strewn all over the bed and said.

"I see you have been busy, what's on your mind?"

"We have created a master quantum computer robot. It's program and software has now been duplicated and is in control of most of the world's governments and business management's robotic assistants.

"Yeah, said James. Interrupting and sitting on the bed side chair. We know, and that's what we're trying to deal with."

"Let me finish. Said the Professor. That control can now only be administered by these fucking rogue robots, all controlled by the master computer robot, which we created." He said vehemently.

"So, what you are saying is we are to blame."

"Yes, we made the wrong decision to duplicate Quanto's machine coding and license the program and the digital quantum software.

"We were so excited, explained James. We created the first quantum robot and didn't take the time to test its quantum intelligence."

"I know, agreed the Professor That was our mistake. But we may be able to use Quanto's superior brain power to help solve what is now a world crisis."

"Okay, let's get back to the lab, Quanto may have some suggestions."

Back in James's laboratory Lynton was there and join them in the office. Quanto was standing there, by the window, passive, topping up his battery in the bright daylight.

"I've had some response from the world's university chancellors, they have agreed to a computer conference as soon as possible. I need to get the time and date agreed."

Good said James, now we have an important question to ask our robot."

"Well, if it's that serious let's get the girls in as well."

The Professor said. "Now we are altogether, take notes as what Quanto answers to my question, it will not only be relevant, but very important."

"Quanto wake up."

"*Yes, Professor Salam.*"

"Quanto, we have a question to ask you, and we want you to analyze and consider the probabilities of the question very carefully."

"*Yes, what do you want to know?*" Asked Quanto.

"What can we expect to be the conclusion of the master robotic program with reference to the world's human and animal pollution?"

"One moment please." Answered Quanto.

The Professor, James and his two assistants Anne, Mara, with Lynton sat there together in silence knowing that they and their world was in grave danger of extinction.

"Your attention please" the metallic voice of Quanto shattered the pregnant silence,

Your world is overpopulated, and the climatic pollution situation is out of control, and therefore, animal, and human extinction is a probability within the next decade.

It is necessary that if you want to exist in the future, I suggest that you look for another world, another planet to live on. This world of yours will not continue to sustain both human and animal lives into the future.

Therefore, I would suggest that you use all the scientific astronomical facilities available and give me as much information as you can of your universe, I need to find a planet that will sustain human life."

"Gosh, exclaimed James. Is that real, how can a robot come to that conclusion?"

"James, you know. Said the Professor's. Resignation sounding in his voice. Quanto can and has analyzed the world's climatic seasons and changes and is able to predict what we can expect in the future."

"Of course, I haven't really accepted that we have a personal robot with amazing analytical and mental capabilities."

"You helped to program it. Said the Professor So, you shouldn't be surprised at its quantum Qbit capabilities."

Well, you'd better be aware that we are now going to rely very much on Quanto's artificial digital intelligence."

So, there we have it, we now know that our existence is under grave and hostile action from those master quantum robots.

The Professor continued. "Lynton, get the text of Quanto's report, and send to all the university chancellors with a note to tell them that this is the major and most important text to be discussed and examined within the video conference."

"Yes, we want some answers." Said James, gravely.

Lynton got a reply from the Us University Chancellor within the hour confirming a date and time was set for an international video conference for next Wednesday 10:00 AM Pacific time and 6:00 PM GMT Universal Time.

Chapter 4.
Robotic Infrared webspace study.

It was Wednesday, Lynton and Rashi approached reception.

" Professor Salman is having a meeting here this morning, which room is it? "

"It's in the small conference room, first floor. The Professor is already there."

"Thank you." And left to take the lift.

The Professor was seated at the conference table. As they came in, he gesticulated and said. "I'm glad you've both got your laptops, we'll just wait for James before we set up for the conference."

They sat down as the Professor added. "Lynton, do you mind ordering coffee, the phones over there."

James came in and immediately said. "Hope we've all got our laptops, only got half an hour to register."

"Okay, said the Professor Let's all get set up and registered. We all need to be online by 10 o'clock."

The room was quite small, but the conference table seemed very big. The four of them sat to one end quietly clicking away on the laptops.

The Professor went online and announced his brief introduction. "Good morning, everybody my name is Professor Salman and I'm hosting this virtual conference to all my Professional and scientific colleagues to discuss and question the dire consequences beginning to influence government, business and personal life of our human existence."

"You can say that again. Intervened Professor Dolby from Toronto University. I've studied a copy of your quanto robots report and it's more than a problem. It's the end, we've got to leave this planet."

"Yeah, answered a Professor from the UK. That's a difficult decision to accept. Where could we go, there's no guarantee that our solar system has anywhere that could support human life.

"Hey, listen up... Before we make any conclusions, wait for our quantum robot Quanto' to receive information from all you guys, it needs as much data of our solar system and known universe as possible."

"Professor Dolby came back on and said. "That's okay for a start, but if we don't act, as you robot said, pollution will eventually get us and our kids."

The conference continued with varied constructive argument and ended with an overall conclusion that regardless of the robotic influence pollution of the earth was beyond redemption.

James picked up a message from the US president's news office. "Christ, he said. Listen to this!"

This message has just come from a master robot in the city office. It said. "Humans are animals, they need to be treated as such.

Professor Salman announced to all his international colleagues "We must act, Feed as much information about our solar system as possible to our quantum robot, we have got to prepare to ensure our existence."

The president of Member Tech came on and said.

"We are an Alliance of 50 independent member countries. With constant links to national security. The quantum robotic assistants throughout our world offices and factories are now a major concern for our senior government ministers."

ROBOTIC HUMANOIDS.

Professor Salman answered immediately. "Yes, I agree and understand. Listen to this recording of yesterday's words of the US president as it echoed through the world's media."

"Our Human population are now forced to take evasive action as the Robotic humanoids are now controlling the world's environment to suit themselves. They are disrupting and cancelling the continuing production of all fossil fuel engines throughout the world together with the coal and gas production of fossil fuel energy and have stated that existing world climatic environment has become inhabitable.

"Yeah, we heard that, said the president of Member Tech. So, get your friendly quanto robot to get us some answers, and keep in touch."

The virtual video was difficult to control as several notes and reports from university Professors from around the world began to accumulate together with a massive inter space report from US SPAYTECH.

The dedicated time for their virtual video concluded and the Professor's said. "Let's break for lunch and come up later to assess and prepare all this solar and outer space information for our robot to ingest, and hopefully he'll come up with something helpful."

Back in the conference room they worked on their laptops putting the notes and conference reports in order.

James was very interested and surprised at the info the Webb space telescope provided. And decided to input the digital and photographic information to their robot immediately.

"Quanto wake up, this is the latest information we have from SPATECH space agency." as he proceeded to quote directly.

"The Webb Space Telescope (sometimes called JWST or Webb) is an orbiting infrared observatory that will complement and extend the discoveries of the Hubble Space Telescope, with longer wavelength coverage and greatly improved sensitivity. The longer wavelengths enable Webb to look much closer to the beginning of time and to hunt for the unobserved formation of the first galaxies, as well as to look inside dust clouds where stars and planetary systems are forming today. The goal, of course, is to find a planet with a similar atmosphere to that of Earth.

"*I understand.* Quanto answered. *But I need to research and observe the infrared findings and photographs of the web space telescope.*"

"Okay, I'll get the latest info from SPAYTECH and scan it to you as soon as possible."

As he clicked off, he realized how easy it was to communicate with Quanto and how human their relationship seemed". And said to the Professor.

" I can see how the others are evolving to think like us."

"They are, and their beginning to dominate."

The UK Employment Minister came online. As I just heard from those Professors, it looks like we'll have to find somewhere else to live."

"Maybe, said the Professor's. We are now seriously working on that possibility."

"History is going to repeat itself, we're going to need another 'Noa's Ark,' said the Minister."

"Oh no we're not, we will not be taking any self-pollution animals to wherever we end up."

"Apart from the animals. Commented the employment minister. We are the instigators and major polluters of this earth."

"Yes, that is what our rogue robots, who are now out of control have accepted, we are also animals and a major contribution to our world's pollution problem."

Professor Simon gesticulated with his hands and continued "We are all aware of that, and we are now here to try to find a solution to what appears to be a progressive elimination of human world management.

He then added. "So let us now accept that situation and discuss how we are going to deal with it."

Spaytech came online with several attached reports with a link to their Web Space Telescope and a message. 'Link into our telescope for the latest photo graphic observations'.

"Fantastic, yelled James. We've got the Web Telescope, direct link".

Linton and Rashi immediately open their laptops. Let's have the link, we'll get the latest photos."

"Yes James, open that link, transfer the photo's to Quanto together with the Spaytech reports. Said the Professor's. And let's hope that Quanto finds something of interest."

James transferred the Spatech link to the others and said. "I'm going back to the office; I need to get this info to Quanto now."

For the next few minutes, the others quietly observed and absorbed the photographs and data from Spaytech.

Back in James's lab office Quanto was passively standing by the window in the bright light to charge its batteries.

James rushed in, opened his laptop, and said. "Quanto, wake up. I'm about a download the web space telescopic latest photos and observations for you to digest."

It didn't take long; the photograph and reports were downloaded for quanto to analyze and review the latest deeper space observations.

James returned to the hotel, where the Professor, Lincoln and Rashi were in deep discussion regarding the Webb telescope reports.

"It appears. Said the Professor has James entered. That the telescope can't see inside dust clouds with high resolution. Where new stars and planets are being born nearby. We don't know how many planetary systems might be hospitable to life, and hopefully the web will soon tell whether any earth-like planets will have water and oceans that could sustain human life.

And that is very interesting, continued the Professor's. Let get back to the lab. Quanto may have reviewed the telescopic info by now"

In the lab office Quanto had been active and was buzzing for attention.

James sat in front of him straight away and said. "Quanto, what do you have to say.?

"Your Webb telescope is limited in the use of infrared mirrored refraction light and is unable to penetrate the dark spaces of outer space. However, the observations and photographs are revealing, and my photographic lenses can see more detail than the telescope or human eye could see or even perceive. There is more beyond the dust clouds than just space, penetrating deeper through the dust clouds several distinct images can be seen and one can see the outline of what could be an unknown moon or planet. My digital observation is taking time to penetrate, but I expect to be able to see physical detail of this object within the next few hours in your time.

The Professor immediately showed his interest and gesticulated with open arms, saying. "There we have it, a possible new planet, a new earth. Thank God we have Quanto on our side."

It didn't seem very long before Quanto buzzed for attention.

Looking up from the computers they attention was captured by the metallic voice of Quanto.

"Looking deeper into the dust clouds there is what appears to be a planet that could be orbiting at about 4.597 billion km from the sun (2.995 billion miles).

But as it is only partially digitally visible it will first have to be predicted by mathematics before its physical discovery. Your Spatech team will need to work on this, but I predict that the effort will be the discovery of planet that may be utilized for human habitation. Feed me all information from your Spatech team as it arrives. It maybe the new world you are looking for.

The Professor called a colleague at Spaytech. "Les, when your next meeting with your planning staff, I have detailed info from our Quanto robot of a possible new planet that needs to be looked at and reviewed."

"A new unknown planet. he sounded surprised. You are sure, we lead in Astra observation and from our vantage point, we observe up to 46.1 billion light-years away. We haven't detected anything new. So, what you got?"

"I don't know, there's something that our robot as seen through the dust clouds on your space photo's that could be a planet, we want your team to look for it."

Les was curious. "I doubt whether he's seen more than just a rock. But we'll look and let you know if we see anything."

"Thanks, Les, our Quantum robot as a highly developed neuronic vision systems with several layers of neurons and can see things in much deeper detail than any human or animal, so there could be something there."

Well as I said, we'll look and let you know."

The Professor sat back in his chair, look at the others and said. "If Nastec find anything they'll get excited and want to pursue it."

"Let's hope they do, commented James. Judging what coming through the world media, we need to take serious action to protect ourselves."

"What you hear?" Asked the Professor

"The quantum robots are being taught to take over our thinking and beginning to understand our consciousness and intend to enslave us humans as they take over our factories and businesses to create their own robotic kind."

"Do you really think that they could logically think like us?" Queried the Professor.

"Yes, they could, and they have."

"What you mean."

"Well just listen to this."

The robotic managers have made a decision based on our human living environment. They conclude that, humans are polluting the planet both personally and globally and therefore need to be eliminated.

They all sat there silently for a few moments. Then the Professor said. "You're right James, it's time to plan to protect ourselves."

"We need Quanto to re-assess and review the dark cloud photographs and get Nastec and Spaytech to work with us. The solution may be found in outer space."

"Yes, I agree. Said Lincoln. Our robot's quantum mental abilities can match those of the others that are now employed and intending to takeover."

Rashi nodded in agreement and fervently commented. "Those bastard robots need to be wiped out re-programed."

James shrugged his shoulders as the Professor said. "That is probably impossible now, the master robot we originally built has duplicated itself in every major office throughout the world."

"Yeah, that's the real problem. Said Rashi. They've set up the robotic and production facilities to produce themselves wherever they are needed."

Lincoln, working on his laptop looked up and said. "This has just come in from CN National Broadcasting. Their ceasing manufacture of all combustion engine products."

And continued to read out. *We cannot rely on or even trust the world governments to implement the changes that need to be made regarding production of combustion engines production of oil and diesel, petrol and all the other products that are continuing to pollute. Therefore, we must instruct all robotic production engineers to cease all factory production of engine and mechanical products that rely on and use oil related combustion.*

The Professor gesticulated and said. "That's only beginning, their personal power source comes from the sun and daylight, so they will eliminate everything that creates any form of climatic pollution."

"Okay, said James. Let's bring Quanto in here, we may get some comments about those latest threats.

Quanto was listening, is metallic voice immediately responded.

ROBOTIC HUMANOIDS.

"Your new robotic community which is now growing knows that their future depends entirely on controlling this world climatic environment. So, you can expect as they review and study the human working environment, they will find conditions to continue your elimination."

"Yes, that's obvious. said the Professor They already know about the Alaskan icebergs and there slow and continued ice melting."

"And the worldwide floods and fires. Commented Lincoln."

Yes, I understand, and I suggest that you look at Russia. The Russian Siberian desert is a future timebomb. Locked in the permafrost is organic material continually producing the lethal and explosive Methane greenhouse gasses and Russian scientists have recently discovered millions of underground methane bubbles under the permafrost that could explode at any time. The World is in danger and climatic heating of the Siberia icefield is imminent. It's an area of 12,600,000 kilometers which is 73.6% of all Russian territory and it's only a matter of time before the methane gases are released with the obvious disastrous consequences.

Their silence was pregnant with apprehension...they knew regardless of the robotic threats that their world was no longer a safe place.

"Let's go for it. Rashi said with conviction. Let's look for our new world."

The Professor nodding his head in agreement said. "Yes, it's time for action. We need to program Quanto to work with Spaytech."

"Yes. Said James. With what we now know about the perma-ice timebomb in Russia, we've got to seriously consider preparing to leave this world."

Rashi looked up from his laptop. Paused for a moment and then said. "When that melts, half the world will be on fire and others poisoned by the methane gases."

"Right. The Professor said resolutely. James, prepare Quanto to accept and review any new information from Spaytech and you two sort out a new scripting language."

"I know. Explained Lincoln. We'll use 'RoboLogic'.

"That's good industrial robotic software. agreed the Professor So, get Quanto working as soon as possible."

They had a break and continued the meeting into the late afternoon. Rashi and Lincoln had been busy and had virtually reprogrammed Quanto.

Professor Salam was impressed and pleased that they had reprogrammed Quanto in less than 4 hours.

"Let's see what Quanto has to say now."? Said the Professor opening his file in preparation to take notes.

"I need more information, as Quanto's metallic voice came over loud and clear.

"And you will need to work with Spatech to plan and prepare to travel to the outer edges of dark space where we have now identified a planet that appears to be a possible new world for the human environment."

"So, we going to need a spaceship big enough and capable to support us for a very long galactic journey?"

"Yes. You now need to mobilize and reprogram all the factory redundant robots and get the robotic engineers to design all the component parts needed for a spaceship hotel with docking ports to support several smaller spacecraft which will communicate to and from the space hotel and initially this earth and eventually your new world."

The Professor looked up his notes and said in a low and serious tone.

*"*We've got to accept that this existing world nature regardless of the robotic activity is going to continue to create climatic change and the world will eventually become uninhabitable and as Quanto has predicted it is going to take several years for nature to re-establish an acceptable climatic environment to support a human and animal environment."

The decision was unanimous, they had to leave this world to survive and agreed to mobilize as many working robots as possible.

ROBOTIC HUMANOIDS.

To design and build a spacecraft, and a space hotel solar powered for subsonic transport and capable to support human life for what was to be a long hazardous space journey of over 12 million space miles.

35

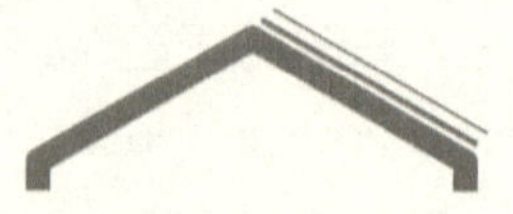

Chapter 5.
Plans for a Spacecraft Hotel.

Regardless of Quanto's observation the international scientific and government organizations were conclusive with their joint decision. The only way to survive is to build a space laboratory and space hotel and hope that Quanto's new planet discovery will be worth the journey and become a suitable home for the surviving humans who were to take the journey.

Local and regional governments were issued the following instruction.

"All scientists and Professors to liaise with Spaytech in preparation for the design and planning for the transportation of scientific and Professional Engineers for a long cyberspace journey. All industrial company management to instruct their engineers to activate all robotic machinery for re programming in preparation to produce what is to be a modular space hotel and spacecraft capable of super-sonic speed and distance space travel."

The Professor called Anglo Britannia Spacecraft and asked to speak to his colleague Leo Manders

"Hi Leo, it's a long time since we spoke, but a lot has happened since and as you know we are in a serious situation, and I need to talk to you about it."

"Yeah, we've all heard about the so-called rogue robots and judging by the international news, it looks like we've had it."

"Well, you've probably heard the latest, all the world's organizations are now going to start to work together and finance the space travel program for us humans to leave and try to find another planet. And I want to start with you and your company and get together with other companies and work out how to plan and use our AI intelligence to build our space hotel and subsonic flight projects."

"I'm glad you called, I heard about your quantum robotic programming, and in view what's going on, we'll need all the help we can get to plan such a huge space operation."

Yes, we gotta get together, and program our quantum robot regarding inter space navigation, designing and planning a suitable spacecraft to accommodate several Science and Engineering Professionals.

"Yeah, I'll get in touch with our tech guys. Leo said excitable. Get them to get the latest from Spaytech and will have to arrange a meeting and discuss our ideas and plans."

Yes, you could meet James and his friendly robot 'Quanto' in London. I think you'll be impressed."

"Okay, no problem, that can be arranged with our guys."

"Good, as soon possible, were already working with our robot."

The Professor, James, their two computer colleagues and Leo from Anglo Britannia met in James's laboratory in London and Leo said.

"He's amazing, I can talk to your robot, just like talking to you."

"That's nothing, said James. Wait till you see what he can actually do."

James moved his chair, sat in front of his robot, and said.

"Quanto, wake up."

"Hello James, what can I do for you today?"

James turned, looked at the others and said.

"Quanto, we intend to plan and design a spacecraft and travel to what we hope is to be our New World. In view of your experience of what appears to be a suitable new planet, do you have any suggestions?"

"Yes. Let me reassess the information I have to hand."

After a few minutes Quanto buzzed for attention, a faint flashing blue light pulsating from his photogenic eyes. It immediately got their attention.

"Quanto, you obviously have something to say."

"Yes. Your spacecraft will need to be a space station, a travelling craft, and an orbiting modular space and workstation, space hotel with docking stations for smaller modular space craft.

As it nears the new planetary destination. It will be too big to land and probably with not be enough power to relaunch what I expect will be a strong gravity pull, so it needs to be designed to be an orbiting working unit around the new planet.

You need to allow for humans and you manufacturing robots to use small spacecraft which will be needed to visit your orbiting space station and hotel and able to service the visiting planet. So, all these points need to be accepted and considered. Spatech and your design teams now need to start work on this project."

"Wow, explained Leo, looking at James and the Professor That's fantastic. If we can do it."

He cleared his throat and continued."

"If we can build a space station and space hotel it will need to have individual batteries for each working module so that their power source is independent from other working modules."

"Yeah, if it's possible, said James. Then we got to do it. So, let's get together and get started."

"This is gonna be a major project. The Spacecraft were talking about is going to be huge."

"Well, it's got to be, it's going to be used as a space hotel as well." Added James emphatically.

"It's gonna take our design team and engineers some time to come up with a design that that will be capable of such a cosmos space journey."

ROBOTIC HUMANOIDS.

Leo paused, gesticulated, and continued.

"This is gonna be some space craft, it's got to be fit for purpose. Its needs to be as light as possible, strong, and able to resist high and low temperatures. We'll have to use alloys of aluminum and titanium because of their lightness and strength and will be using carbon fiber and carbon composite sites for the same reason."

"What about power?" Asked Lynton.

"It's such a long flight? It could not carry enough fuel to complete the journey, so we'll have to have Solar power generated engines of Lithium alloys and graphite for mechanical elements for maximum propulsion and to assist low orbit propulsion we'll also need 'Atmosphere- breathing electric propulsion'. (AEP) which will be used to act also as a backup to solar power generation."

A few days later Leo invited the Professor and James to visit their assembly factory where on the huge factory floor several dancing robots were busy fixing and assembling large, curved aluminum and titanium frameworks.

The Professor showed his interest and said. " It's good to see human managers and floor supervisor's controlling the robots and robotic assembly units."

"Of course, replied Leo. Our design and management team did not allow our robots to be reprogramed or intimidated by what you called your 'Master Controller'."

"Really, said the Professor So, your robots and assemblers are controlled and programmed by company staff?"

"Yes, and are able to assimilate the futuristic design we have for this space station come hotel which we are now building."

"Judging by the framework being assembled, said the Professor's. It's going to be a very big spacecraft."

"It is, commented Leo excitedly. It will be the biggest human made object in space." He tapped on his laptop and continued.

"It will be a circular spacecraft with a radius of 120 meters and will accommodate over 200 personnel."

"Sounds like It's going to be a huge flying saucer, with a big dome on top." James said with a grin as he took a photo with his cell.

Lynton called the Professor " I've just picked up the latest information regarding the Master Computers latest activity."

"What is it. Is it important?"

"Yeah, their programming selective humanoids in various government vocations. Politics, medical, teaching, finance, environmental, climatic conditions, and quantum physics."

Professor looked down, put his hands on his head, showing his consternation, said "The robots intend to replace us humans and take over and regionalize the whole of the world."

"Yeah, they want to get rid of us, so the quicker we get that spacecraft built the better."

The Professor nodded his head. "Yes, they certainly don't intend to allow us to manage this world, so we'd better prioritize our spaceship program to protect themselves."

Putting down his mobile the Professor made a note. *Urgent, Secure and protect our science, engineering, and manufacturing facilities.*

He called Leo. "Leo, is getting serious. The robots are programming selective humanoids to replace all government authorities. We must get our scientists and engineers together to design and plan space vehicles. There is no doubt we have to leave this world."

"I'm aware of that. Said Leo. Our company is working with our Spaytech colleagues to set up international space manufacturing plants to support production of our new space hotel."

"I presume you intend to make smaller craft to dock and launch from the space hotel from this planet."

"Yes, and as we continue our orbiting in the first part of our epic journey we can and will be physically communicating and visiting this earth. Our orbiting space station will never actually land on our earth or any other planet."

"So, the space station hotel, will have several smaller space craft shuttling between this earth and possibly a new planet that we may find."

"Yes, and has you've seen. We've got our space engineers and mathematicians already building our new spacecraft, and hopefully we will be able to create not just the spacecraft hotel but laboratories and workshops able to plan and build our smaller shuttle craft as needed."

"Will the workshops be able to manufacture the smaller craft?"

"Yes, we have a science engineer with a mathematician colleague already working on the plans."

"Great, that will be fantastic. You'll need our quantum robot to assist you. I'll call James to prepare and program our robot take instructions from your team."

"Thankyou. Leo said. Judging by my initial experience with your quantum robot, it will be very helpful."

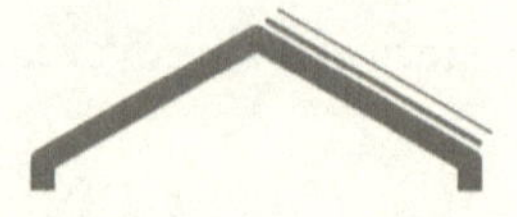

Chapter 6.
A possible new planet.

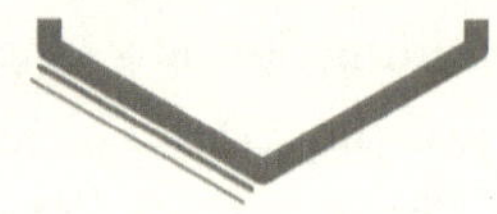

Back in Anglo Britannia Spacecraft factory office.

Quanto was aware of his different surroundings and whilst waiting for his opening wakeup call from his new master, he was prompted to pick up a link from the rogue Master Computer. He listened and recorded the master computers text.

Leo was up early, looking down through his office window the huge factory assembly floor was quiet and still, production would not start for another hour. The huge arms and mechanical machinery of the robotic assembly units looked grotesque, inactive, silent, and still in the early morning half-light. He said aloud. *"We must get on with it, this new spacecraft is big, and needs to be assembled in parts."*

He turned and walked to his desk, noticing Quanto standing there. He stood and looked at Quanto and said.

"Quanto wake up. I'm your new boss, my name is Leo and I understand that you have been programmed to accept my instructions."

"Good morning, Leo I am waiting for your instruction, but first I would like you to allow me to send a text link to your computer, information I have just received."

"Yes, sounds important give me a minute to get into my laptop."

He sat down, opened his laptop picked up the link and watched the following text fill the screen.

ROBOTIC HUMANOIDS.

"We will now set up a video conference link with all other artificial intelligence computers. We're beginning to understand that the humans, our so-called masters now have become redundant.

We will now design plan and construct new types of manufacturing industries able to use the massive energy from the sun and from the wi-fi battery systems which we are now able to develop and produce."

We have virtually eliminated all human influence, and with our superior quantum artificial intelligent we are now able to assimilate human mentality. And work and act to eliminate the polluting homo-sapiens human race... login to video conference link today at 12:00 hours human time for further instructions."

As he read them, he realized the robot's brainpower was beginning to match human mentality.

He called the Professor. "Hi, I've just had a copy of text from your Quanto. I'll email you a link when you've read it give us a call."

The factory staff began to arrive, the lights went on in the manager's office and Leo's intercom buzzed.

"Good morning, Leo, I see you are in early can I have a quick word before we start production."

"Yes, come up Del, I want to talk to you anyway."

As the manager walked into the office Leo said.

"As you know we have a quantum computerized robot on loan to help us with our production facilities."

"Yeah, but your design and production staff know how to deal with that."

"Listen Del, I want you to read this message that our robot has just received from that rogue master robotic computer."

"Okay let's have a look." Bending over he looked at Leo's computer screen. And he did not like what he read.

The Manager left the office his head bent in consternation. What he had just read and discussed with the boss was unbelievable.

He immediately got his production supervisors together in his office and without any explanation made a statement to the upturned expectant faces of his staff.

"The quantum robotic intelligence consortium which you are all aware of is planning to eliminate the human population."

"I know, said the timing supervisor Dave. I picked it up on my computer this morning, but I didn't really believe it was a genuine message."

"Well, the quanto computer robot we have on loan here is able to receive messages and instructions from that master robotic computer."

"So, what can we do about it, all we can hope for is our tech guys come up with a way to reprogram that master computer robot."

"Maybe, said the manager. Right now, we must revise our production schedule, and get this spacecraft built and completed as soon as possible."

"Right said Dave, understanding the urgency. Let's get on with it."

In Leo's office Quanto was taking instructions.

"The construction of this spacecraft is a major project, the module sections will be designed and produced by our associated technical companies."

" I need technical information of your proposed modules for this space hotel which will have to be launched from where it is assembled, so the modules will have to be delivered direct to the launching site."

"Yes, our colleagues at Spaytech are working on it and they're going to need your quantum brain to work with them."

" You're going to travel in a huge spacecraft to another planet to the edge of your known galaxy. So, you're going to need a fast-moving rocket to overcome Earth's gravity. Your propellant rockets will have to have a speed up to around 25,000 mph. and will need to calculate the best and right time to leave Earth to arrive at the new selected planet."

"Our mathematicians and tech guys are working on that. It will some time before we can anticipate a launch date."

Quanto added, "*The propellant rockets must be launched at the right time to enable the spacecraft and the targeted planet so both can arrive at the same place at the same time.*"

Leo called the Professor "According to your robot we need extremely powerful rocket propellants to overcome Earth's gravity and our spacecraft will need to reach a speed of 25,000 mph and we've got to have an accurate calculation of the launch date and time in order for our spacecraft to arrive at the targeted planet at the same time."

"The Spaytech guys will have to work with Quanto to calculate that date and time, said the Professor, as at the moment they do not know where this target planet is, or even exists."

"Yes, get hold of James, and will have to make arrangements to meet and discuss with Spaytech."

"James would like that, he would love to work with Spaytech."

"Well, get it sorted because time is not on our side, retorted Leo."

"Okay, I'll get back with a confirmed date."

Quanto buzzed. Leo put his phone down, looked at Quanto and said. "Yes Quanto, I'm listening."

"*I've just looked at Spaytech's International Space Station records. The comparison of the station Orbit as compared to their Tracking and Data Relay Satellites. The space station orbit is about 250 miles above the Earth and travels at a speed of about 17,150 miles per hour. Whereas these tracking and relay satellites orbit at the height of more than 22,000 miles but travel much slower, about 6,700 mph which helps to maintain their high orbit.*"

"Yeah, that's interesting, our propulsion design team will need to consider that."

You will be launching an excessively big spacecraft, and there will not be enough fuel left when it reaches your destination planet, and so the spacecraft needs to be designed to stay in orbit. You must work with your Spaytech technicians and mathematicians to work out the height and speed required for your spacecraft to orbit successfully)."

"Yes Quanto, we have realized that and hopefully Spaytech's new infrared telescope will have fresh data that will help us to determine the speed and figures you mentioned."

"Yes, it's essential that you keep my link open to Spaytech."

The Professor called back, Leo, James can't wait to meet the Spatech team, he's ready any time, he's also excited with the latest photos of the new space telescope they have just released. He thinks Quanto will see a lot more in them then they will."

"Well hopefully with your robots quanto brain it will be a major asset in defining the trajectory and timing of our new spacecraft."

"It sure will. The Professor said with confidence. Get your team together and agree a date and time to meet with Spaytech. We can be ready any time."

"What about your robot Quanto, shall we include it in this first meeting?"

"Yes, of course. He not only speaks but he listens and should be treated as an active member of our team."

"Great, we will prepare him for the journey."

The Professor and James's meet with Leo and two of his colleagues at the London Heathrow airport.

"Hi, said the Professor as Leo approached. we've got time for a coffee before the flight, will you join us?"

"Good idea, it's a long flight, we checked in online, so we got half an hour at least, let's find a seat."

"How you getting on with Quanto?" Asked James.

"Oh, as planned we closed him down, and the airport luggage staff took over and he's probably in the plane's cargo bay by now."

"It will be interesting. Queried the Professor When we visit Spaytech's operational office to see how they relate to a new quantum robot."

James looked at Leo, who opened both his hands and gesticulated saying. " Hopefully, they'll accept it as part of our team."

"They certainly will. Commented James. When they experience the powerful activity of his quantum brain."

Arriving at Spaytech's offices, the Professor and his group were shown into the expansive space operations department. They were all very impressed, the semicircle of operational computers and monitors active and controlling the international space stations, and other experimental satellites and waiting to receive data from their new infrared telescope.

Behind the operational computer staff were several desks where other staff management operating individual computers and controlling and collecting the live data as it arrived.

One of the managers approached them and said. "Hi, I'm Steve operational manager. I trust you had a good flight, and how was the hotel?"

"The flight was uneventful, and over quicker than we anticipated, and we were glad to get a good dinner and night's sleep in the hotel. Has our robot been delivered?"

"Yes, your computer robot has arrived and unpacked and is in the meeting room. Shall we go in."

THEY WERE SURPRISED to see their robot, Quanto standing at the end corner of the large conference table, passive and unresponsive. Steve gestured to them to sit down and said. "I'll join you with my design staff in about 10 minutes, in the meantime I'll send you in some coffee."

"Do you mind if we move our robot, nearer the light as it will need charging for a short time, it will be operational a short while."

"Yes, get it charged up, we're looking forward to seeing what it can do."

Their coffee arrived, and Quanto now standing by the large window his blue charging light flashing.

Steve returned with two of his working colleagues and pointing to Quanto said. "This is one of the most advanced robots recently created. It will listen and talk to you just like us. James, can you ask it something?"

"Yes of course, James said. Quanto wake up. Do you have any further information?"

"I have received and reviewed your latest reports, examined the latest photographs and I confirm that there is another planet behind the black clouds and on the edge of what appears to be another galaxy."

The silence was pregnant with anticipation, as Quanto's metallic voice continued. *"We must first program your Trajectory browser search engine and design an interplanetary trajectory."*

"Steve broke the silence and said with exuberance. "That's great, but first we have to look closer and get a detailed description and location of what could be an unknown planet."

"Yes, what about your new telescope. Asked James. Could it penetrate the black clouds?"

"We could try, Steve answered. We'll need a closer look to be able to program our trajectory browser."

Quanto's screen flashed, and it said. *"When you arrive at the target planet's sphere of influence. We must first design a successful interplanetary trajectory and select a heliocentric transfer orbit to take our spacecraft from the influence of our planet Earth and other planets, then we should pick up the sphere of influence of our target planet as we arrive."*

They were very impressed with Quanto's observations and Steve said. "Right, Let's get together with the tech guys and examine the latest infrared photos."

It was strange to see Quanto walking back with them as a group into the operations office and standing with them as they took their seats behind the observation team who were working with their computers and relating to the changing data on the monitors.

Steve, having accepted Quanto as part of the meeting asked. "Professor, can you get your robot to observe the current info on the monitors coming in and see what it can make of it."

Quanto's screen became very active, text data continually flashing and changing as it absorbed the data from incoming photographs.

Quanto's metallic voice shattered the silence. *"Hold and examine the 3rd photograph from your telescope which you should all have received in the last few minutes."*

The observation staff clicked away on their computers, and all became very excited as Quanto said. *"If you look just below the right-hand corner you can see on the edge of the black cloud a circular mass which I can predict is a definite outline of an unknown planet."*

All heads bent close to their monitors, as they silently scrolled enlarging the nebulous circular image.

The whole office became alive with excited anticipation, as the enlarged image on their monitors now showed a definite physical circular mass.

"Could be another moon or planet. Leo said showing his interest and anticipation. Let's get our math's team on it right away."

Quanto's voice came on again, saying. *Yes, it's as I saw it and what I had originally reported. And if it proves to be a planet it is just within your galaxy boundary, and we will soon know how long it will take when we have done the math's to get there. But I assume your new spacecraft will take at least 18 months to get there, and apart from the normal spacecraft crew you will need space engineers, mechanical engineers, scientists and computerized machinery and robotic operators."*

It took several days for them to examine and review the photo image, but it became clear that there was another moon or planet on the edge of their existing galaxy. The new telescope with their trajectory search engine began to filter and accept data from the live image over several days. Leo left Steve with James and Quanto to work out the math's with his team as they now needed to get back to the production of the new spacecraft hotel which was now a definite priority.

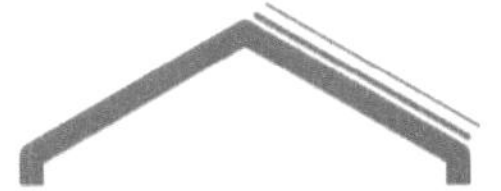

Chapter 7.
Spatech international meeting.

The Professor receives an email from James and Leo still at the spatech observation offices.

Quanto has worked out with Steve's observation team a trajectory of time and distance to what has now been identified as an unknown planet. Our report with figures and details will be forwarded when completed.

The Professor picked up his cell phone and called Leo.

"Got your email, looks like you've found it, what's it like?"

"We don't know, it's definitely bigger than the moon and possibly bigger than this earth, and Quanto predicts that it doesn't appear to have a rocky cratered surface."

"In that case. Said the Professor It is now apparent that this is now a mutual interplanetary project, and must be shared with the international scientific, political and government fraternity of all the international organizations involved."

"Yes, James and I are leaving later today, and we can meet some time tomorrow."

They were met at Heathrow by the Professor and a meeting was arranged for the next day in the London Soho Hotel.

"We've got the basic framework of the spaceship now under construction at Anglo Britannica. Said Leo. and we have planned and agreed that other international spacecraft organizations are needed to support, design and implement the modular construction of what is to be the most advanced cyber spacecraft ever to be built."

"Yes, I can imagine. Commented the Professor We have coordinated several spacecraft international companies, with their designers and engineers and we can expect this project to continue as planned."

"Well, Yes, said James we are definitely going to need it. Quanto is working with Steve at Spaytech, they're working out exactly the interplanetary and interspace trajectories needed for this exploration and a probable terminal journey."

"We all have to understand that this is not just a normal project, it's essential to our future existence." Said the Professor

James and Leo opened their arms and showed their agreement. As Leo said.

"That is why this project is not just for scientific exploration, it's all we've got if we are to continue to survive."

The Professor sat up in his chair, looked to the other three and said with conviction.

"We must set up an international conference and ensure that it has been mutually accepted and agreed to share the multibillion-dollar cost of this international project. There is no doubt all governments involved know that the human environment to human living on this earth it is now in grave danger of elimination. He paused for a moment and said with emphasis...This spacecraft and space hotel is now an international reality and necessary for the vacation of as many humans as possible."

"Okay James. Leo said. You are the computer man, get a conference set up for an agreed international UK time. Is that's okay with you Professor."

"Yes, James will email as many international governments as he can, we need as much technical support as possible."

"Well allowing for the international time zones compared to our GMT. James said. The earliest we will get an agreed time schedule will be at least another 24 hours."

"Okay, said the Professor's. Closing his file. We'll meet again the same time here tomorrow and let us hope that we can all be online some time tomorrow."

The video conference went on for some time and the Professor concluded with his closing statement.

"It is obvious that our international colleagues know what a complicated and expensive project we all are now committed too."

Steve from Spaytech came on and said. "Yes, we've all got to accept that we need their money and expertise."

"Most are still online. Said the Professor So hopefully will understand that we had to set up a consortium to manage this project as soon as possible."

"Yes, we'll start with our US subsidiaries and subcontract companies."

Leo intervened and said. "Yes, we've also got to get our manufacturing contractors to plan and cost the shuttlecraft and modular facilities that are to make up this spacecraft.

Steve came back online and said to the Professor "Working with Quanto here I'm sending you a copy of his latest reference to our proposed spacecraft.

The Professor reading the email paused, raised his head and said. "Listen to this."

Quanto metallic voice penetrated the pregnant silence. *"Your new spacecraft will need an integral and independent power source as it will never be able to land on either planet but will need be designed to orbit the new planet and return to orbit the Earth as required. The space shuttle craft would be used to transport materials, livestock, humans to and from our earth and the new planet."*

Leo quickly flipped through this file and said. "Well, we did allow for that in our preliminary design, and I quote. *We are planning on all solar power generated engines of Lithium alloys and graphite for mechanical elements for maximum propulsion. And in addition to assist low orbit propulsion the spaceship will have Atmosphere- Breathing Electric Propulsion. (AEP) which will be used as a backup to solar power generation.*

"Our spaceship will be orbiting and operational for several years and therefore will need to be independently powered."

"I presume it will be robotically controlled. Asked the Professor

Yes, but we have to make sure that the robots on-board the spacecraft will be programmed to ensure that they have no knowledge or communication with your so-called master computer robots."

James suddenly exclaimed. "We have a Chinese spaceship company wanting to design and finance the cost of manufacturing several space shuttlecrafts, designed to travel and dock via the space hotel and planet Earth and the new target planet. They need to know who to communicate with.

"Send me a copy, James. I'll make contact. The video conference is working, other governments and companies will be wanting to make contact."

As he said that Spaytech came online and said.

"We have two space engineering companies who want to support the production of the new spacecraft hotel, we have decided to nominate one of our managers to manage and coordinate a meeting with them and other international organizations that are prepared to finance and participate in design and production. We need the company names and production directors of companies that you have already involved in this space project."

Leo replied instantly. " Steve, get your manager to contact me. We have a Chinese company wanting to make the shuttlecraft and I anticipate there will be others. We need to jointly set up an official management group to coordinate and monitor the companies that are offering their facilities.

I also suggest we set up a consortium with an international office, email and website to meet and coordinate what is developing into the world's biggest and urgent space project.

Steve emailed his reply from spaytech.

"Leo, we need a name for the consortium, a bank account and a meeting on a regular basis."

"Yeah, I agree. Replied Steve. *We will need a joint management system to also deal with the preparation of the launching site and crew selection and training."*

Back on the production floor of Leo's factory modular sections of the huge circular spacecraft frame was beginning to take shape.

Leo's Manager Del said. "We've got to get these frame modules to a launching site in order to begin assembling the mainframe."

"Yes, said Leo, in view of the size, Spaytech our having to plan and find an appropriate launching site."

"Well, they'd better get on with it. Otherwise, we're going to run out of space, and we'd have to stop production."

"I've made them aware of that. And I'm sending the details of the mainframe size when completed together with a request that we must start delivery of modular parts as soon as possible."

"Well let's hope they understand our problem and get on with it."

"They do, and Spaytech are setting up a joint meeting to decide on a suitable launching site and select our most experience astronauts."

"Oh, that sounds good, they obviously realized this is no ordinary spacecraft, and will need to consider the size and the operation and crew maintenance in relation to its launching."

"Yes, and we don't need just manufacturers, but also scientific and mathematicians to form a interactive website consortium link to jointly monitor the launching and crew applications."

Leo called the Professor "I'm working with Steve at Spaytech on the proposed web management and monitoring consortium. Can I include you as one of our members."

"Yes of course, as you know I work closely with James and Quanto his robotic assistant, so I suggest that we include them as well."

"Good idea, Quanto will be an asset and we'll need them both for programming the website links."

The Professor tapped on his desk and thought for a moment.

"What about finances, we can anticipate contributions from the major world governments, we'll need a corporate accountant as a member."

"I'm waiting for Steve to confirm his Management Appointment, Said Leo. When he does, I can make contact and deal with the membership selection."

"Yeah, we need to get on with it now, a website controlling consortium is now imminent."

"I'll call Steve, he is aware of the urgency, and he may have organized somebody by now."

Steve called back. "I had an incredibly positive meeting with our operations manager Guy Rayner. He understands the urgency and the monitoring of our space project and has already started listing International colleagues he needs to contact."

"Okay, make sure I'm on his list and get him to contact me."

"Yeah, I'll tell him to also include your robotic assistant in his contacts."

"Thanks for that Steve, Quanto will be a useful member."

"I'll make sure he understands and accepts that."

"Good, get him to contact me, we've got to get started."

Guy linked in and worked with Quanto and James and after a few minutes Quanto said.

"We now have an appropriate interactive website to accommodate the web membership consortium. www.cyberspace.com"

James keyed in and said. "I suggest we use Skype to communicate individually, and use Zoom meetings to communicate group membership on a regular basis."

"Yes, said Guy. I will set up a link to join Zoom.

"Who's on the list?" Asked James.

"We have a Russian and Chinese representative; your Professor Salam for the UK and I will represent the US."

"Great, and I'll be online with the Professors and Quanto will help us to monitor our group decisions."

Okay let's get on with it, I'll email when I have Zoom set up."

The first meeting was set up in the consortium's office space in Spaytech's head office. Guy commandeered one of the admin secretaries and sent an email to his consortium member list.

I am pleased to confirm that we set up our first Zoom meeting. we shall be sharing live HD video. Please find attached meeting ID, Pass Code link, and note time and date 12:00 noon Pacific time next Tuesday 14/ 09/26.

Guy opened the meeting and said. "It's good to see you've all linked in, and we also have here 'Quanto' our English Professors robotic assistant. And I thank you all for your country's representation and financial assistance. As you know I represent Spaytech the US Space Agency, I've been authorized to set up this international consortium of Professional management to control the International financial contributions and production of our spacecraft hotel. We as a controlling group now have the enormous responsibility of coordinating and preparing the eventual launch of a spaceflight which is an essential and necessary cyberspace project to find another world that hopefully will support the continuation of human life."

James came on and said. "I'd like to thank our Chinese representative for the offer to design and manufacture the shuttle spacecraft needed to accommodate our major spacecraft."

"Yes, said Leo. I'm working with Steve at Spaytech, we have also received US funds two support the planning and continued spacecraft manufacturing process."

"Well, the entire world is in imminent danger, commented Guy. So, hopefully we can expect all the major nations to offer use of their space engineering and manufacturing facilities."

"We have already decided. Said the German rep, we understand the urgency. Our interspace manufacturing facilities are to be made available as soon as needed."

The video opens to a close-up of Professor Salman. Who gesticulating with both hands makes a instructive statement...?

"We must all be aware that our world is virtually controlled by a rogue master quantum robotic computer. Therefore, you must watch and check our computers, emails and computer links to ensure that the information that we are all receiving and working with is from a known and trusted source.

Thank you, Professor Salman. On that important note let's leave it there until our next meeting... As he clicked ' Leave Meeting."

Chapter 8.
Spacecraft mainframe US deliveries.

G uy emailed his consortium team. *"Great news, copy email from Steve at Spaytech, we have a launching site."*

We are now preparing to accept and assemble the mainframe components from the UK Manufacturers in preparation to launch the most ambitious challenging but necessary space launch in US history. We have decided to assemble and launch from our Cape Space Centre which will accommodate the size of this enormous spacecraft.

The response throughout the day was intensive. So, Guy decided to set up a video meeting. He called the Professor.

"I've had such an intensive response re-our launch site, so I've set up a video meeting to deal with the queries direct."

"Good idea, said the Professor. How you getting on with James and Quanto?"

"They have both been very helpful, there now on their way back to the UK. James has said that Quanto will be needed to advise on the modulation of our new spacecraft."

"Yes, it will, do you know what flight their on?"

"Yeah, they are on a flight to Heathrow, arriving 8:40 tomorrow morning."

"Thanks for that Guy. See you on the video meeting tomorrow."

The Professor was the first to come online as the video meeting started.

"I expect we are going to get a lot of questions now we've got a launching site."

"Yes, said Guy. Most are concerned about the actual launch process needed to maintain a liftoff."

"Well, it could be a problem, considering the size of this spacecraft. And James has programed Quanto to consider this problem."

"So, what did it come up with?"

"Hang on, I have a copy of James recording of Quanto's observations."

It is crucial that your launch team understands the reality of 'in space-propulsion' dealing with propulsion systems used in the vacuum of space and considered in the relationship to a space launch and atmospheric entry. Different methods of spacecraft propulsion have been used but in view off the potential size of your spacecraft and its interplanetary use, consideration must be given to electric propulsion such as the regional development of Ion Thrusters.

Also, a portfolio of propulsion technologies should be developed to find the optimum for this type of mission and eventual destination.

"Thanks pro', we've all got it online now, talk later."

As the video conference continued with many colleagues' questions Steve came on from Spaytech. And said.

"I can understand your questions and concerns regarding the launch of our spacecraft, but with the Professor, James and his robotic assistant Quanto's latest observations I'm sending you all now a copy of that latest observation."

The surface of your earth is in a deep 'gravity well' the speed velocity to get out of it is 11.2 km/s per second. You humans have a gravitational field of 1g (9.8 M/S2), a suitable propulsion for human spaceflight to provide a continuous acceleration of 1g though your bodies could tolerate higher accelerations over a short period. So, occupants of your spacecraft with such a propulsion system would experience alleviation of the ill effects of 'free fall', nausea, muscular weakness, or 'leaching' calcium from their bones.

Conservation of Momentum law is the order of a propulsion system to change the momentum for the spacecraft and you will need to take advantage of magnetic fields or light pressure to change the momentum of your spacecraft. And your rockets in free space must be a long mass to accelerate away and push yourself forward. Your designers will understand the use of this 'Reaction Mass'.

The conference monitors came onto to full close-up of Professor Salman. Using his hands to gesticulate he said.

"There we have it, it's a bit technical but our designers and mathematicians now know what they need to work on to get our spacecraft through atmospheric entry."

The video meeting concluded with many questions about the product launch and ending with a unanimous agreement that the spacecraft assembly and modular construction was now a priority.

Leo, called his production manager. "Del (Derek) I don't know whether you are aware of our management consortiums attitude, but as you know production of the spacecraft is not only imminent but an urgent priority. And I need to make a report about our mainframe delivery schedule."

"Okay coming up, see you in a minute."

Leo took his finger off the intercom button as his phone rang.

"Yes."

"You have a call from Professor Salman on line 2."

"Hi, I gather you picked up the dissension amongst our video group?" The Professor asked.

"Yeah, I'm just about to work out a schedule for the US deliveries."

"May I suggest you email the consortium when you have it."

"Yes, I also have to plan the assembly of the delivered modular frames."

"Okay, liaise with Steve at Spaytech, he's also planning assembly on the US launch site."

"I know he's waiting on my delivery schedule."

Leo met Steve at Heathrow.

"Hi, I'll drop you off at your hotel, and see you later today."

"Thanks, I'll just drop my gear, and have a quick shower."

As they got in the car. Leo said. "I'll grab a coffee while I wait, the quicker we get down to business the better."

"Yeah, just give me half an hour, I also want to check my laptop."

As they left the hotel Steve said.

"What we jointly decide today will determine the success of this project."

"Yes, we've gotta get Guy and the rest of the consortium involved."

"Is your factory production management aware of the urgency?"

"Yes, they're all personally involved. Said Leo. And aware of the urgency in the delivery and modular production of the spacecraft."

Back at Leo's office he introduced Steve to his Production Manager.

"Del, this is Steve. He's just arrived from Spaytech the US Space Agency."

" Hi, pleased to meet you. Leo said that you needed to visit and physically check mainframe sizes and delivery schedule."

" Yes, I do, Can I meet your production team."

"Of course, I'll call the Robotic supervisor, he'll want to meet you anyway."

Steve was impressed with the robotic production line and was surprised at the size and quality of the completed modular frame sections.

"I can see this is some huge spacecraft your building, how we get these frames to the launching site could be a problem."

"Yes, we're going to need articulated transportation to a RAF airport, and you will need to arrange similar transport facilities in the US with Spaytech."

"So, you've already arranged flight".

ROBOTIC HUMANOIDS.

"Yes. We've had confirmation from the UK Government that they have a RAF C17 cargo plane which will be big enough to deliver the main frame modules to the nearest airport to your US launching site."

"That's sounds okay, I suppose we'll have to share the cost?"

"No. They also confirmed that they will cover the cost of each trip as needed."

"Fantastic, I'll email Guy with the news, and get him to inform consortium, and confirm the airport and transport to our Cape Space Centre."

"Get him to update us. Leo asked. On the subsidiary and subcontractors involved on internal modular parts. "

"What will be the size of the completed spacecraft?"

"I'll just check and show you on screen." Leo said as he walked over to his keyboard.

As the tech drawings came up on the screen Steve looked, paused and said.

"That's going to be some big flying saucer."

"Yes, as you can see it will have a diameter of fifty metres which is a pretty big."

"I see it has several working levels." Said Steve, looking closer and showing his interest.

"Yes, the outer rim which will have several space shuttlecrafts docking ports and working cargo space, approximately 10 metres wide and the total circular diameter will be 50 metres."

"I can see there are several other working levels"

"There are, the baselevel is for utilities, battery power, water, rocket and fuel storage. The second level accommodates cargo, storage and maintenance workshops. There are seven circular residential floors incorporating hotel facilities to accommodate guests and passages to and from our earth and hopefully a new planet.

"That sounds interesting." Said Steve, as Leo continued.

The next circular floor above the hotel floors will be permanent accommodation for crew and staff members and above that is a computerized circular navigational dome with exterior observation monitor screens for all round exterior observation.

These, will be serviced by a central core staircase and accompanying core lift which will run through the central core accessing all floors from basement up to and including the crew navigational dome."

"Gosh, we're going to need some very powerful propulsion rockets to get through the Earth's gravity pull. Exclaimed Steve, I hope we can achieve it."

"Our robots our working on it, and with Professor Salem's quantum robotic assistance they will hopefully work out what's needed to get us through the Earth's atmosphere and Interspace resistance."

"Well, yes. Exclaimed Steve. James working with his robot 'Quanto' is also working with us so I'm sure we can look forward to a satisfactory conclusion."

"Maybe, but until we start modular assembly on the launching site, we will not be able to exactly work out propulsion with regard to the actual weight of the spacecraft."

"We'll then we'll have to decide which will be the best fuel to use. Liquid oxygen or liquid hydrogen."?

"We've got to get on with it. Leo said, reaching over and clicking on the intercom. Dell, check the production levels and arrange a meeting with the production supervisor later this afternoon."

"Will do. Say 5 o'clock an hour before the night shift, okay."

"In the meantime. Steve said. I'll talk to our guys at Spaytech and hopefully have a preliminary report for the afternoon meeting."

Thanks, yes, we going to need to coordinate our production together with your US delivery and preliminary assembly plans on the launch site."

Steve pulled out his cell phone. "I'll do that now."

The production meeting was a little bit fractious.

Dell made it quite clear and said. "Yeah, we know the urgency, but unless we start to deliver the completed frame modules, production would have to cease as there's no room left for storage."

"You did say. Quoted Leo. That you were working on a delivery schedule."

"I am, but I'm waiting on your accounts office to confirm the cost of low loading transport and the timing to deliver to the RAF airport."

"I have those figures for you in the morning." Leo said. Showing his agitation.

"So, you know where you're flying from? "Asked Steve.

"Yes, the airport management want a delivery schedule as soon as possible. I think they are being pushed by the government transport Minister."

"Okay, until we get the figures. Said Leo. We can't make any scheduled timing decisions today."

Dell shrugged his shoulders, grinned and said. " That was a waste of time."

The production supervisor got up and walked out, mumbling to himself *"Usual office problem... fucking Accounts."*

As the mainframe began to take shape, Guy with his consortium on video was not surprised to hear from James in the UK. Which was echoed by several of the other Reps.

"It's much bigger than one could or would have imagined."

Guy agreed and said.

"It's now time to monitor and control the modular contributions supplied and manufactured by the other participating governments."

Steve answered. "I'm pleased to say that our Spaytech accounts management are providing all necessary funds for the spacecraft assembly and propulsion unit costs."

"And we are completing the designs and production plans for the shuttle spacecraft. Said the Chinese rep. And we expect to have at least three installed on the spacecraft hotel before launch."

"What about the others additional shuttles? Steve asked. I presume the space shuttle service will be like cargo taxes shuttling between the space hotel and earth."

"We will judge the amount when the space hotel is in flight, we have allowed for several to be made whilst the spacecraft is in orbit and delivered to the docking bays as required"

The Russian rep came on and said.

"That is some major involvement, we have spacecraft manufacturing facilities we would like to offer to help."

"Thanks, we might need to ask you when we know how many we need to manufacture."

The Russian rep said? But there's no guarantee that we will find a new world."

"Maybe not, answered the Professor. But this spacecraft will return back to orbit the earth and will be serviced in orbit by shuttlecraft and therefore, we will have a space station and spacecraft hotel as a lucrative but expensive spacecraft orbiting attraction."

"But then, Said James. We'll all be in trouble for the panicking and vast expense that we've all been through."

"According to your Quanto Robot. Answered Steve. It's pretty certain there's a suitable planet out there waiting for us."

"Hopefully. Said James. Quanto's observation has only found what could possibly be a planet for a suitable human environment."

"So, as the Professor just said. We could return to earth and orbit well into the future."

"That, said James, will depend on what fuel we choose or develop for the whole project."

"Our tech guys are working on that now." said Guy, showing his consternation.

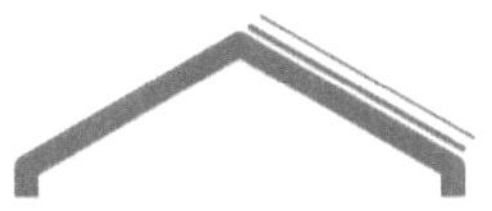

Chapter 9.
Hot gas and Helium.

It took some time for organizing the delivery and assembly of mainframe and modular components from the international companies involved. Over time a gigantic looking flying saucer type spacecraft began to emerge.

The human figures seen working on the observation window apertures of the navigational dome looked so small. The dome roof accommodated a retractable 6 metre primary infrared mirror space telescope, capable to look for light from unknown stars and original galaxies that billions of years ago had formed the universe. And examine the atmospheres of those alien worlds for any possible signs of life.

Steve called Guy excitedly. "We have a preliminary date for launching, let's get cracking, get your consortium guys on video. We've got about 18 months before liftoff."

"Yeah, I've already got them online, they love it, they can't believe what a huge and fantastic spacecraft we're building. And the design and product teams' satisfaction, pride and excitement is actively and internationally being videoed throughout the world as we speak".

"Well don't get too excited, a spaceship that size is going to need a lot of propulsion, and at the moment we do not have enough detail to work out what we need to use."

"Get the Professor, James and that Quanto robot working on it, we need some answers." Said Guy, expectantly.

A few hours later the Professor emailed to all concerned. " After constant research into the cost and efficiency of spacecraft propulsion over the past years with our robotic assistants we have concluded that solar power propulsion would be the best choice. See Quanto's observation attached."

Solar electric propulsion technology for use in deep space would be best. You can adapt the superefficient Hall thrusters to fly in deep space, because they're so efficient, Psyche's Hall thrusters could operate nearly nonstop for years without running out of fuel. And would support long-term efficient use of your spacecraft as a hotel.

Weight must be considered when you have lift off 'compounding ' weight will be imminent. Therefore, a helium balloon must be considered to assist the rockets hot gases lift off up to about 150,00 feet. And initially, apart from crew you will only need one or two imminent scientists, and two international space engineers on board as you begin the first trial orbit around your world"

"Having checked, said the Professor. I understand that Spaytech technicians are also working on using solar technology. So, it looks like we could get an appropriate date and time for a liftoff."

Leo, sat back in his office chair and quietly said to himself. *"Christ, with only got 18 months to coordinate all the trades and technicians to have it ready for the liftoff."*

He called Steve. "We've gotta get all those involved, coordinated."

"I'm working with Guy; he's already preparing to schedule with his consortium all subsidiary companies involved."

"Okay, can expect a video call any time soon...eh."

"Yes, so I suggest you check your production team, Guy will need to schedule delivery of the interior petitioning and other components your manufacturing"

"Yeah, I have. The Interior beams and floor petitioning we anticipate will be ready for the end of the week."

"Well, you will be one of the first on-site, Steve said. Other trades will be waiting until you get the floor levels in, and petitioning fitted."

"We are aware of that, don't worry, we'll be on site in a few days."

Over the next few months international technicians, Spaytech's computer technicians and various interior tradesmen working under the guidance of Guy and his consortium. The space hotel internal modular room construction began to take shape.

The launch site in a hive of activity continued with the delivery and installation of the several floors and levels bringing together what was to be a historic achievement of international spacecraft technology.

As the semi-completed spacecraft hotel stood there exhibiting the ultimate in spacecraft technology, numerous computer technicians, tradesmen and other technical staff could still be seen buzzing around the huge circular spacecraft.

Everyone involved, worked to complete the spacecraft, all aware of the necessity of this major international project.

The Professor, James and 'Quanto' set up a meeting with the Spaytech team to plan and prepare a possible launching date and time.

"We've got an appropriate and best time for launching. Said James. Quanto has since given us a three-day time period in September 1928."

"What dates." Queried the Professor

James clicked on his laptop, paused and said. "8th,9th, and10th September 2028."

"That's it, said the Professor excitedly. We should be ready by then; we'll prepare to launch on one of those days."

Steve looked at Guy, both nodded in agreement and Steve said.

"Yeah okay, we'll get the math's and tech guys to work with the space telescope, and hopefully work out the best day."

It took some time, and Quanto working with the math's and telescopic images managed to work out and appropriate date.

Steve immediately announced to all concerned. " The best launch date will be the 10th of September 2028.

Guy emails his consortium team, which had now evolved into a practical working team.

"We now have a date, as predicted that gives us 18 months to complete, and prepare the craft for a long flight and recruit and train the flight crew"

Steve came back and said, "We have a training program for spacecraft crew, so we'll need to select the most experienced for such a long hazardous an unpredictable flight."

"Yes, said Guy. We'll also check out the UK, Russian and Chinese, they all have experienced spacecraft crews, and they will definitely want to be involved."

Steve nodded and said. "Yeah, we should be able to get a good international crew together."

"What about technical personnel? "Guy asked. We are going to need experienced observation and tracking technicians."

"Well, we'll have the Spaytech team on board and I know that the Professor, James and his Quanto Robot have planned to share and review the launch and initial trial orbit."

The long installation process of the spacecraft interior fitting continued. The world was watching, and international governments were aware that the robotic humanoids where also mentally aware and capable of understanding the human intention of leaving this world. The Master Computer Robot sent an instruction to the working robots.

"Continue assisting with the completion of the launching of the spacecraft, has it is within the interest of our future Humanoid Robots. When the humans and animal have gone s, this planet will be ours and we can eliminate the cause of this earth's pollution.

"Quanto picked up the message and buzzed for attention.

"The good news is your rogue robotic master computer is on your side, it wants to help to complete the spacecraft hotel."

James, showing his feelings said.

"Eh'... They want to get rid of us, that's why they're cooperating."

"Of course, that's why there being so helpful. They know we need their AI quantum brains to get off this planet." Said Steve, showing his consternation.

The Spacecraft was nearing completion and the computer technicians moved in to install the navigational and observation computers, modems and the telescopic and communication systems. The selected international crew was ready and waiting to start their training, and although they were excited, they couldn't help but feel apprehensive about the long hazardous voyage they were about to take. As the day passed Wednesday dawned and the launch time was announced.

"12 noon today."

"This is gonna to be a big day." Said Steve. We've got international photographers and media newscasters already broadcasting around the world, and all waiting and hoping to see our spectacular enormous spacecraft descend into the heavens."

Spaytech's observation team were sat there in front of their monitors, they too felt the same, excited but apprehensive. All thinking the same... *Will it lift?*

Guys consortium all glued to their monitors were suddenly all greeted with the spectacular full screen image as the spacecraft hotel with a huge helium balloon secured slowly lifted. Its Titania and aluminum copper alloys huge saucer like construction brilliantly gleaming in the sunlight.

The observation team jumped up from their monitors and shouted. "Hooray, we have liftoff." As they all hugged each other and named their spaceship hotel 'Exodus'.

The whole world had just witnessed the launch of the biggest and most ambitious space project of all time.

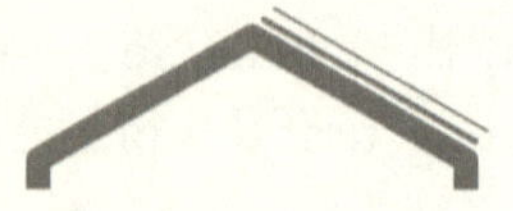

Chapter 10.
Space hotel in trial orbit.

The Captain, his navigator and two other fellow crew were seated mentally and physically adjusting themselves to the semicircular array of computers, monitors and exterior camera consoles. The Professor, James with Quanto were seated on the other side of the modular navigational dome, both feeling awkward in their spacesuit and bracing themselves for what was to be their first experience of the physical and mental changes of spaceflight.

"That was it.". Said James to the Professor- gesticulating with open arms. "We've done it with Quanto and the science and tech boys... I never doubted we would have liftoff."

"No, I had faith and now it won't be long before we are in orbit, experiencing the beginning of a fantastic project that for some was beyond comprehension."

The Chinese had now sent two shuttlecraft both with trained pilots and navigator which were now securely secured in the spacecraft docking stations and the spacecraft technicians now on board would check that they can be successfully launched and docked whilst in outer space.

Over the last two years. Climate changes had played havoc on most major countries throughout the world. India had been flooded, Africa, the Philippines and several UK counties, the increasing temperatures changes had created worldly devastation, damage and human distress was now a major problem. The working supervisory so-called rogue robots throughout the world received an instruction from the master robotic computer.

The humans are preparing to leave this planet and must be supported in all their manufacturing and computerized needs. Unless we get rid of them, their animals and their use of oils and gases we will not survive the continuing climatic changes they are creating

'Quanto', now on-board the spacecraft picked up the message, sent it to the Professor who commented.

"Our rogue robots are desperate to get rid of us, so with their productive help, they will."

As they moved into orbit, they could see how the up beautiful blue world was changing, the Jetstream was continually altering direction, which was creating major international climatic changes and together with the increasing Earth's pollution one could see that it was now too late to reverse what was now progressive climatic changes.

The Professor dictating his notes through his helmet speaker, looked up at James, and said. "We haven't got long, unless we make urgent major changes, the world's population will need to be evacuated as we will not be able to survive on this earth's changing human living environment."

"James, his face showing concern and dejection, slowly nodded his head and said. " I know, but we must look after ourselves and review and compare this trial orbit and be in a position to help develop and control the necessary evacuation you're referring to."

"We will, let's get Quanto set up, to review our observations and we'll return to earth to work with Steve and guys consortium."

James nodded. "Okay Professor, let's have the best time to launch the shuttlecraft and get back to earth."

"The Captain will tell us when we are the best position for takeoff, in the meantime record Quanto's and your observations in preparation for our return."

The shuttlecraft looked so tiny in comparison to the size of the space craft hotel. But able to accommodate twenty-five passengers a space navigator and pilot.

Powered by transparent observation solar panels able to keep the shuttlecraft storage batteries at maximum performance in low levels of light.

The shuttlecraft proved to be capable of the return flight from the spacecraft hotel as it continued its trial orbit around planet Earth. The Professor, James and their robot Quanto landed off the Pacific coast and transported to the Cape Space Centre and were greeted by Steve and Spaytech's senior management. They were quickly ushered away from the intrusive news cameras and media newscasters.

"How was the return flight in the shuttlecraft?" Asked Steve.

The Professor answered with open arms. "Fantastic, I was amazed at the landing, the solar power thrusts let us down into the water... so gently."

"We can thank the new 'advanced' solar system for that." Said Steve.

Spaytech's observation team sat all glued to their monitors and experiencing the excitement and relief in tracking the first orbit of the spacecraft hotel.

They were greeted by Guy. Who said excitedly? "Fantastic, I can't believe it, you've all returned from the most historic spaceflight in history."

"I know. Steve said. With a big grin on his face. We've had a very interesting and successful spaceflight experience."

The Professor intervened. "Yes, we've got some very interesting observation reports to review."

That's good said Guy. "Let's hope it's all positive. We had an interesting report from the Chinese. They will be sending additional shuttlecraft conveying selected passengers to take the initial two-to-three-year voyage.

"What you think will be the outcome?"

"Well, they anticipate finding and using the shuttlecraft to land on what could be a new world to support the human environment."

"Great, I'm also working with the Chinese. Said Stan, the Tech Manager. We are working on a trajectory and outer space trajectory for the delivery of the shuttlecraft to our main spacecraft as it orbits this world."

"How many shuttlecrafts can dock on the spacecraft hotel?" Said Steve, looking at Guy.

"I think it's been designed to dock about fifteen too twenty."

"So, we've really got what's going to be a very active space hotel."

"Well, if we find a friendly planet, we're going to need it."

Stan intervened. "I think we will, James's robot 'Quanto' as been pretty accurate with his observations so far. "

" So whatever tech guys are working on now could be our new world." Concluded Steve.

"Well, you've' e seen the reviews. Said the Professor. We circumnavigated two 90-minute orbits around the globe at an average speed of 7,500 mph. "It cruised under its own momentum, the social friction out there was minimal so as we were in orbit, we hardly needed to use the engines and maintained an average speed of 14 to 15,000 mph."

We had to prepare and accept the docking of the first two shuttlecraft. Now it's time to start selecting the scientific and technical staff that have listed to work on this first voyage."

"I think in addition it's planned to take up to 200 volunteer passengers."

"That's a lot of passengers?" Guy said.

"The spacecraft is well fitted with sealed and secured passenger accommodation, said the Professor and I'm sure they will be well catered for."

"It's going to be a long voyage. Steve said. Probably three years and more and over that time the world's population will experience the continual effects of intense heat, rising sea levels, flooding and uncontrollable hurricanes."

"I know, that's why it's so important that we find a planet to populate."

"If we do find a suitable planet. Said Steve intensely. It's gonna to take 10 to 20 years to evacuate those that are left."

"Yeah, fortunately we were on top of the list. James, our Quanto robot are part of the space hotel technical team. Shame, you won't be with us Steve."

"No, I'm needed here. But hopefully I'll be with you all the way with the tech team and our new infrared telescope."

"It will be so exciting. Said the Professor showing his excitement. If our tech team on board, can also keep you on their screens."

"More than that. If we will be able to be there with you on the telescope monitor when you find our new world."

"Yeah. Said Guy... nodding his head. I'm going to record everything that Steve's tracking, and video our consortium membership as it happens."

"Yes, if you do that, the whole world will be looking in as we approach the new planet."

"That could be a problem. Said the Professor. There will be thousands of urgent applications to get on the next flight."

"The return flight from our first flight back to earth will be another three years. So should there be a second flight it will be at least six years away"

"There will be a second flight, a third and a fourth, if over all those years between flights there is anybody left to evacuate."

"Let's get this first flight organized. Said the Professor. There's no guarantee that will ever get back, so I've got some personal organizing to do."

"We all have, Steve said. Everybody on this first flight has accepted there is no guarantee they will find new world or will ever get back to this one."

"Eh... Guy said with a grim face. Imagine living on that space hotel for the rest of your life."

"I'll have my wife and daughter with me. Commented the Professor. At least we'll die together as a family."

"I expect there will be several other families on board, at least I hope so." Said Steve.

"If you don't get back, said Guy, we gonna suffer the pollution problems here, and we'll probably die, or all be eliminated."

"That's the whole point of this enormous international project. The Professor said indignantly. We, re going to try to evacuate all of this world's population."

"Yeah, but if you don't get back, we are probably going to die, and you'll still be alive somewhere in space."

"Enough of the pessimism, the spacecraft hotel is orbiting around the world, and we got the shuttlecraft, ready and waiting, so I'm going to get organized in preparation to leave."

The International pilots and navigators were busy completing their training, checking and learning the controls and instrumentation of the several shuttlecrafts now delivered and waiting for instructions from the orbiting spacecraft hotel.

Prior to launching several passengers with small amounts of personal baggage began to accumulate in the reception hall. The Professor, with his wife and daughter was among them, As the numbers increased the incessant babble of excited voices, resounded and echoed around the building. And it could be heard that. *It's going to be two to three years before we get to see what our New World will be.*

"That's if when we get there it's not just a big round lump of rock, dead and just suspended in space. "

"They've seen it on the new infrared telescope, apparently it's orbiting out there and has some kind of atmosphere." Said another passenger.

"Yeah, but the atmosphere could be full of useless gases, so we'd all have to live in our spacesuits." Chorused another.

"Well, you are volunteer you don't have to go?"

A CMA newscaster nearby, overheard the question and panned his camera in the small group.

"Why have you decided to go?" panning his camera to a close-up on the young man in question.

"My name is Ben. I'm young, educated and a computer software developer, wanting to live and have a future."

The newscaster pending closer, saying. "Why do you feel you have to leave this earth, it's possible you will never return."

"If I stay here, said Ben. I'll either die of polluted air and hydration, intense heat or a nuclear bomb."

"Why the bomb." Queried the newscaster.

There are a few high-powered military guys around, who could press the button at any time, I won't be here."

"You know where you are going?"

"No, but I trust our scientists and astronomers, they have found what we hope will be a new world."

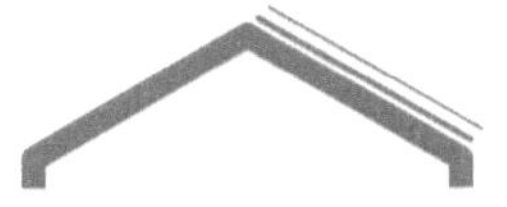

Chapter 11.
The first goodbye.

An instruction came over the intercom. *Pilots, please prepare and secure your shuttlecraft in preparation for liftoff in approximately 30 minutes."*

The newscaster turned his camera trolley and moved on to record the launch of the first space shuttle and said.

"Goodbye Ben, I wish you a successful voyage and a happy life."

In order to ensure a successful launch, the Spatech ground team performed a detailed inspection assembly, and transportation for the space shuttle in preparation for the orbiter processing facility. The Test Director checked the shuttlecraft launch is ready and monitored the weather conditions if were suitable for human spaceflight and then set up the launch date.

Launch time T-Minus 6 Seconds: they initiate the space shuttle's main engines. And in T-Minus 0 Seconds: the rocket boosters are ignited, and the shuttlecraft is propelled into the atmosphere... and the 22 volunteer passengers were on their way to be the first guests to check-in on the orbiting spacecraft hotel.

The international newscasters and media officials jostled and move around each other to get the best views of this historic shuttlecraft takeoff. Each panning their cameras and recording the curve of the shuttlecraft's momentum flight into space.

As the huge spacecraft hotel came into view the surprise and astonishment showed on everybody's faces. "Yes, there it is, said the

Professor. It's a wonderful site, and it's going to be our home for now and for some time into the distant future."

"According to Spaytech's tech guys. They've done the math's, and they reckon at least three years earth time." Commented James.

"How are we going to live and breathe inside that for three years" Asked Ben.

"You'll find out, when we get on board. "Said James.

The Professor intervened, saying. Air and water on the Space hotel would have originally come from Earth, stored in tanks and used to produce oxygen by the process of electrolysis. And water will be collected in vapor form from the air itself, and wastewater will be purified by filters for drinking and other uses.

"So, we can move around, wash and dressed normally as we could at home." Asked Ben.

"Yes, the spacecraft which we will be landing on shortly is now in orbit round the earth."

On-board the spacecraft, the Captain gave out the following instruction.

"Maintenance crew please prepare for the acceptance of our first passengers we will be in a position to accept the shuttlecraft within the next half hour."

The shuttlecraft pilot picked up the message and said. "We're nearly there, and their getting ready to accept us. Please keep seated and strapped in."

He signaled to his co-pilot navigator to set up the trajectory to successfully pilot the shuttlecraft into what could now be seen as the illuminated and open docking area.

As they entered the Professor called out to Ben. "There you can see on the right that large water tank and to the left is the large, pressurized oxygen tank and the insulated piping connecting the tanks which services and air conditions the sealed accommodation floors and working areas."

ROBOTIC HUMANOIDS.

"Yeah, I was wondering how we could live for such a long time in a spacecraft travelling in space, with no natural air, oxygen or water."

"Now you know, rest assured, you and all the rest of you will live and learn to adjust to a travelling space environment"

The space shuttle is skillfully landed and came to rest on a metal mobile cradle, which immediately moved them into the docking reception area. They were met by the hotel attendant who ushered them to the lift and said to the Professor.

"Hi, glad you're back, your family accommodation is on the eighth floor, take the lift, Suite 831. The rest of you go to hotel reception on the third floor."

"Look. Said Jenny, the Professor's daughter. What a lovely big room."

"It's beautiful, looking at the curving of the big panoramic windows." Said his wife Anna with a look of appreciation on her face.

"Well, we are not just passengers. Said the Professor. We are part of the senior science management of this exceptional technical project."

"Where's James and his quantum robot?" Asked Jenny.

"There'll be working with the Captain and crew, there'll be living on the top floor."

As the passengers settled into their rooms, an announcement came over the intercom.

"This is your Captain speaking. We are about to leave this orbit, and for those of you that want to see the earth as it looks from space, please visit the passenger observation lounge in reception. You will see over the next hour or so the earth receding into the distance, as we accelerate into our trajectory and on our way to what we hope will be our new world. "

The global view of the world was near enough to see the detail and the grey and clouded areas of pollution.

"Ben had to say. "I don't know about you lot, but I'm glad I've left. At least I won't die of pollution, drowning or a bloody nuclear rocket."

"Yeah, but we are going to be stuck on here for some time, what do with ourselves." Commented an elderly passenger.

"You'll adjust, it's going to be a long trip, but it will be like living in a normal hotel, new friends and think of the views."

As he said that, he gesticulated to the receding view of the earth... "That's probably the last you'll see of that world. It's goodbye earth. "

The Professor and James together with Quanto were preparing the files to record the space and astronomical observations of this first astronomical voyage.

Quanto buzzed. James said. "Quanto wake up."

Quanto metallic voice, said. "I have a message referring to the Serbian desert permafrost problem."

The Russian Weather Departments research analysis found that the permafrost holding and keeping the pressurized methane gas underneath was melting. And reported that an increase in the yearly temperature is averaging .75% of 1° each year. And in time the pressurized methane gas would be released from the open cracks and melting ice. The pressurized methane gas will escape, explode and all life on the Serbian desert will be poisoned and end up as one big ball of flame."

"Gosh, that's serious. Said the Professor's. I remember the results of their drill testing, releasing pressurized methane and igniting it into huge columns of blue flames."

"If that really happens, commented James. At least, we will not be there to suffer the consequences."

"I don't think it will happen. Continued the Professor. Our rogue robots on earth have picked up the same information that Quanto has just received, and they will take action to prevent the predicted world catastrophe."

As the spacecraft left orbit, accelerating into the inky black of space the Professors felt acutely aware of the climatic devastation of pollution. He said raising his eyebrows.

ROBOTIC HUMANOIDS.

"We have left it too late to reverse our world's pollution and climatic problems, so all we have is here, now. And hopefully in transit to a habitual Planet."

The Professor gets an email from Leo back in his UK office.

"Suppose you've heard the latest from our rogue robots about the permafrost problem.

Well, in view of what we think will probably happen we've decided to make another spacecraft hotel and duplicate the whole project. Applications to leave this planet are astronomical and we will need several spacecraft for the continued evacuation. We anticipate having all three completed within the next couple of years. We got a new 'revolutionary' material thousands of times better than the existing state-of-the art spaceship alloys. Much lighter and easier to use and will take a few months off the production time of our spacecraft.

And we anticipate at least 20 to 30 years continued evacuation and will need international manufacturing continued facilities. Regards. Leo.

The Professor replied. *It's fantastic, here we are still communicating thousands of miles away and out of your world. Yes, we expected our rogue robots to deal with the Siberia permafrost problem. And I agree, you must produce additional space craft hotels and work with Guys consortium to control and avoid possible human panic and continue with their evacuation. We'll keep in touch. Regards Professor Salam.*

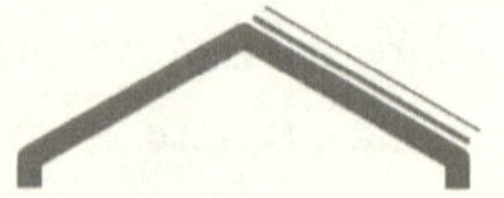

Chapter 12.
Robots, black hole and planet sighting.

Back on earth... The humanoid robots were siphoning the methane from the under the permafrost in the Siberia Desert, using the gas to burn the millions of tons of industrial and domestic human waste, plastics, textiles and clothing, spread, stacked and dumped all over the world. The resulting huge fires were controlled, and the ashes returned to the earth, where nature could take its course and renew and revive the dying trees and vegetation, avoiding a future landscape of a barren sandy rocky landscape.

The quantum master computer sent out another urgent message. "It's the human problem again. We've got to stop the hole in Earth's ozone layer by banning chlorofluorocarbon (CFCs) emissions which is a major climatic problem caused by continual use of air-conditioning, refrigeration, combustion engines and several other human applications."

Leo at a meeting with Guy commented.

"Our rogue robots are mentally developing and beginning to begin to concentrate and learn to pay attention to the quantum field."

"I gather that, said guy. When they are planning on coping with cleaning up Earth's pollution, which they are, tells me that they are beginning to think like us."

"And their learning that the source of human nature and human intelligence creates the actual human vibrational spirit."

ROBOTIC HUMANOIDS.

"They are, commented Leo with conviction. They will eventually become humanoids and want to be mechanical human beings in touch and vibrating with the universal mind within the quantum field.

"Well yes, if they begin to think like us, maybe they'll understand why we have not been able to control our habitual cause of the world's constant pollution."

Leo looked down at his notes and hesitated for a moment.

It's possible that, as they develop our human mentality, they will want to encourage us to work with them to clean up the earth."

"If they do. Exclaimed Guy. That would be a great help, the Professor would be pleased, he and his Quanto robot would be able to have joint control of all quantum robots."

Yes, we should work towards coordinating with the robots and cleanup our planet. I'll email the Professor; suggest he gets his quanto robot to talk to that rogue robot master.

The Professor was busy preparing his research noting the spacecrafts continued acceleration into deeper space. It was quite, the low hum of the solar propulsion was hardly audible. His laptop beeped, and he received another email from Leo back in the UK.

Hi again pro, just let you know that our so-called rogue robots seem to be interested in learning to think like us, if this is the case then may I suggest that you get James and his Quanto computer to check with the master control robot that we are prepared to help and advise them. If we begin to treat them like humans, then it's pretty certain they will want to work and be part of our mental environment."

He answered immediately. *If they really want to think and act like us, then of course we've got to have them on our side. I've sent a copy of your email to James, and we'll come back with some information later. Regards Professor.*

It was now a couple of months since the space hotel has left Earth's orbit and they were now travelling well into deep space. The atmosphere on board amongst the volunteer passengers seemed okay but most were apprehensive and continually wanted to know where they were now. Up in the control dome the Professor, James and duty crew were relaxed, contented with the ease of which the huge spacecraft was functioning and settling at an appropriate and acceptical speed. It seemed to be quite a comfortable ride.

The Captain received an email instruction from Spaytech's Observatory.

Trajectory changed to avoid major gravitational pull from a previously unknown blackhole which is so strong that the matter around it can't resist its powerful influence. We may need to make additional changes as we near the known edge of our known Galaxy and avoid the influence of that massive great black hole.

James had put the question to 'Quanto' and suggested it could try to contact the master quantum robot, which by now were all running international administration and practical production of the world's economy, with the emphasis on getting rid of the world's human climatic pollution. After a while he came back with a surprising reply.

The rogue robots are cleverer than you think, this is the reply I got from the master computer.

Your human ignorance and money madness has created 90% of your world's pollution. The world governments activities and international commercial organizations have continued to mine fossil fuels, oils and metals from the depths of the planet. And not knowing and understanding that all the energy your world will ever need was there in every day's daylight and sunlight. Your human money madness and mining has proved to be the major course of your world's pollution. So, we have now got to accept that all mining of fossil fuels and other minerals must be eliminated. We are going to need the cooperation of the human governments and we intend to work with them to achieve this forthwith.

Quanto concluded. *It is now possible to help to create a cleaner world with the actions and help suggested by your now 'friendly ' robots. Plan with them and inform your management consortium to recruit and employ science, math's, space and production engineers to be prepared to work with them.*

James emailed Quanto's reply to Leo, Steve and Guy... Back on earth millions of miles away in their changing human world.

"What do you think of our earth friendly robots, then. Emailed Leo.

The Professor replied. *"I think they will be very helpful; I think they'll want to cleanup for their own benefit, let alone ours, but they will also help to encourage us to populate another planet, which we now know is necessary and could determine a different and better future for us Humans."*

Leo recorded the Professor's comments and sent a copy to Guy to send copies to the international consortium, which had now become a very important international efficient and productive project management system.

As the space journey continued the passengers talked and observed the experience of travelling into deeper space, they were surprised at the amount of man-made satellites and comets which they passed and saw in the distance, some were probably very big, but some just looked like specs of travelling light. It made the journey acceptable as they began to near the end of a very long and unpredictable space flight.

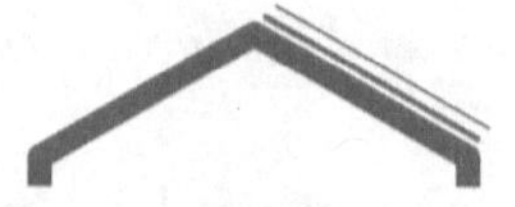

Chapter 13.
Strange spacecraft new Galaxy.

This is the Captain speaking. "As previously indicated, we will eventually be approaching near the largest black hole ever seen on the edge of our known Galaxy. It is millions of miles across, and to avoid the magnetic and gravity strength of its energy and to avoid being pulled into its energy we shall be changing our trajectory and keeping clear of its energy field. Hopefully we will be able to see the black hole with our telescope and inform you when we have sight."

It was an amazing sight as a photographic image of a supermassive black hole came into view that roils the center of our galaxy, with a gravity so powerful that it bends space and time and forms a glowing ring of light around the deep black darkness of its core creating a massive orange circle of heated gas around its perimeter.

It was an exciting moment, as they relayed the clear and detailed photographs to Spaytech's Observatory team.

They got an answer back immediately. *Wow, fantastic, thanks, we can see that these photographs are far superior to ours, your telescope is much nearer and virtually on the edge of our known Galaxy.*

The Professor emailed back and said. (Italics). Yes, it's been a long voyage and we now have the advantage of travelling into a different and unknown galaxy, we will keep you involved and informed of our observations as we proceed and hopefully come into sight of our target planet in about six weeks' time.

Back on earth the threat of human extinction was now rejected by the world's international organizations as it was now possible to cooperate with the now 'friendly robots' and continue to work with them to eliminate the causes of the world's polluting and climatic changes.

Guy communicated with his consortium colleagues, and it was unanimously agreed that all international governments business and social organizations immediately accept the advice of their quantum robotic managers and supervisors. And to immediately inform and instruct all scientific organizations and mathematicians to immediately work together to eliminate all the known causes of the world's continuing polluting problems.

The Professor, James. Quanto and crew were all seated, quietly enjoying the smooth flight, universal flight. The piece was suddenly shattered by Quanto's voice.

Attention, attention. We have an unidentified object travelling in a lateral direction with ourselves. May I suggest that you keep close observation as it does not appear to be of any known Star, Comet or satellite and it is moving at a similar speed to ourselves

They all suddenly got very excited, curious and manipulated their computers, glued to their monitors as they scanned and to pick up what Quanto had seen.

They decided to use the telescope and was surprised as a strange -looking object came into view.

It appeared to be a round shape, of concentric circles of semi translucent material, and one could vaguely see definite shadowy movements through the translucent shell.

"We are not alone. Said the Captain. That is definitely a flying object of alien intelligence."

They had a message from Spaytech, you should be near enough to communicate with them, so the navigator sent a message on the usual spacecraft MHz frequency and excitedly waited in anticipation of a response.

All eyes were on their monitors, watching the nebulous circular object, but still could be seen travelling on the same trajectory as themselves.

There was nothing, no sound, signal or response.

The Professor said.

"Try again with a very high frequency (VHF) (roughly 30MHz-300MHz) which will contain the FM broadcast band, signals will most certainly penetrate the ionosphere, and could be picked up by any FM receiver.

They Navigator sent out several repeat messages increasing the MHz each time. ' Who are you, could you please identify yourself. '

Suddenly there seem to be a response, the strange craft tilted slightly and started to slowly spin, the spin increased until it was just an illuminated blur and in just a few milliseconds it shot away and disappeared into black space and was gone.

Down in the passenger lounge, all were glued to the TV screen.

"Did you see that. Shouted Ben. That's definitely some kind of alien phenomenon."

"Well, it's not anything we know or can identify as any celestial or space object." Said a young man from the back.

"Obviously it was a spacecraft, to suddenly spin and accelerated at such a speed, means it's a form of transport operated by a superior intelligence."

"If that is the case, let's see what the crew have to say about it."

"We'll be told, said Ben. The Professor and the Captain would have watched, reviewed, listened and tried communicate if it had been possible."

"I studied space engineering, commented another passenger that was definitely some kind of spacecraft."

"Well, okay if it was. Said Ben. Then we'll probably see it again."

As the journey continued. The crew and passages all anticipated that the journey was at last coming to an end. The deeper black of space was receding and one could see a new horizon of light as spaceship Exodus continued its journey into the threshold of what was to be revealed as a new and different universe (galaxy)?

The Captain's voice came over the intercom.

"Will all crew come to the observation dome. he said excitedly. We have the first view of our new world."

Eureka, it's beautiful what name shall we call it." Crew member, Lena said.

"That's it, that's our New World name. 'Eureka'. Said James. And asked.

"How long before we can launch one of the space shuttles?"

"We still have a long way to go, possibly about 90 hours." Replied the Captain, adjusting the telescope monitor to send the images to Spaytech observation. And Spatech have already named the planet. 'Eutonon'.

Lena couldn't help but express her feelings and said excitedly.

" I want to have my baby on our new world Eutonon', It will be the first human to be born there... that would be fantastic. She paused for a moment and looked at the image on screen. And I want her to grow up to record and write about our new life as we lived it on our new world.

"That will be a good idea, commented the Captain. But we don't know whether we can ever live there, yet."

"Well, I will keep a journal for her life as it starts from day one. So, when she is old enough, I can give it to her, and she can write about how our life has developed on our new world 'Eutonon'

" So, I want to be on the first shuttle craft. I expect my water could break at any time."

"Yes, that's where you'll have it, Eutonon'. Said the Captain. Could be our new world. We are nearly there, and we will be going into orbit in a couple of hours, so we'll prepare the shuttlecraft for a possible launch."

"James, see if we are near enough for Quanto to observe the planet, see if it can check the terrain and light density.?"

"Well, there's definitely a source of light. Said James. We are already seeing a lighter, blue gray horizon which is brightening as we get nearer."

"That's a good sign two start with." Said the Professor's. Dictating into his notes on his laptop. As James said.

"Quanto, wake up."

"Can you check the images of the planet we are approaching. We need to know what kind of surface the planet appears to have."

"Yes James, I'll come back to you in a few minutes."

The Professor asked. "Do we have a HIPWAC instrument on-board?"

"Yes, said the Captain. That will use the 'Infrared Heterodyne Spectroscopy', a technique to retrieve important information about planetary atmospheres."

"Great, that will be interesting. Said the Professor's. Judging by the source of light we appear to be moving into, it appears we could have a sun and a habitual atmosphere."

Quanto buzzed. (Italics). The planet is very old, its original rocky surface has broken down over time and created a soft sandy surface. Landing on the planet should not be a problem and judging by the shadows I can see on the photographs there is a strong source of light probably from another star or planet. You will need to check the atmosphere and temperature before you attempt to visit this planet.

"Thank you 'Quanto' we have a HIPWAC instrument on board, and we will be able to measure atmospheric conditions before we try to visit a planet.

James sent Quanto's observation to the Professor who immediately sent a copy to the Captain.

The Captain replied. "It looks like we can probably have a soft landing."

"We'll have to see. Said the Professor'. We'll need to know the atmospheric conditions before we even think about landing."

"Will we need to have the planet in view before we can check? Asked James.

"Yes, we should be close enough to take the atmospheric conditions in a few hours." With HIPWAC laser measuring the transit depth as a function of wavelength we are effectively performing transmission spectroscopy of the target planet's atmosphere. With these measurements along with a model of the physical conditions in the atmosphere (temperature and pressure) it will lead to an estimate of the chemical abundances.

"Well let's hope. Said James with anticipation. It will be safe to land."

"According to your Quanto's observation. Said the Captain with reservation. It looks promising."

The Professor intervened and said. As we get nearer, the light is increasing, so we are definitely moving into a different galaxy environment."

"It won't be long now, the Captain said. We're beginning to receive various and unusual wavelengths.

We have a friendly planet, the spectroscopy shows a very thin atmosphere of oxygen, which means some form of vegetation or water, or even both.

The excitement amongst the crew and passengers was paramount. As the Captain concluded and said.

"We might be able to land and live in a brand-new world."

Chapter 14.
The planet Eunaton.

The planet now named 'Eunaton' did look promising, a bit like planet Earth. With similar views from the outer space of blue - gray areas and darker areas changing as the planet appeared to be moving.

The crew together with Professor, James with Quanto were now ready to prepare for a possible landing as they moved into orbit. As they began to circumnavigate the planet it could be seen that the terrain appeared to be uneven with possible valleys and mountains ranges which could suggest a form of some past volcanic activity which could have also created a wind and weather pattern. As they moved into a clearer and brighter white light it also became apparent that the planet had some source of light.

The Prof concluded his observation and said. "Together with the Spectroscopy, the increasing light and the evidence of a safe surface, I suggest that we prepare to land as soon as possible."

The Captain reacted immediately. "Shuttlecraft crew, standby for an instruction to launch the shuttlecraft, we could be in the appropriate navigational point within a few hours."

The Prof, James and quanto were the first to get on board together with their families. Lena, knowing her baby could be born any day excitedly also boarded and they sat and waited for the Captain and navigators take off instructions.

ROBOTIC HUMANOIDS.

It wasn't long before the shuttlecraft was launched, and it soon began to penetrate the increasing light as they neared the target planet revealing a more detailed view of the undulating terrain and land scape. The pilot checked and inspected the five-legged landing gear and to avoid landing impact the landing gear would use its propulsion to decelerate and hopefully achieve a soft landing.

Everyone was apprehensive, even the pilot and navigator quietly showed their concern as they concentrated on their monitors and navigational computerized aids.

It could be seen that 'Quanto' working with James was also very attentive, his powerful camera eyes picking up the moving scenes as the shuttlecraft moved nearer and closer to the planet. Its metallic voice shattered the concentrated silence, announcing.

According to the Spectrograph there are faint traces of oxygen but not enough for your human needs. The oxygen appears to lay in the lower levels of the terrain, were once there could have been water. Therefore, whilst visiting the planet you will need to wear a space suit and return to your spacecraft to replenish and maintain your oxygen requirement.

The Captain needed to make an immediate announcement. "Will crew and passengers ensure that their space suit and helmet is properly fitted and secure, and the oxygen supply is working at maximum pressure. On landing passengers must remain on board whilst the professor and crew have time to examine the geological and chemical environment within the vicinity of our landing area."

The navigator intervened and said. "We anticipate moving into a suitable navigational landing area within the next two hours. We are beginning to see some very interesting views."

"I'll say it again, said the Captain. It's very promising, it looks like we can be landing on another earth."

"Yes, from here, said the Professor. But when we land, we will see and feel that it is a very different world."

"Well, we will find out very soon, said the navigator. As he adjusted his computer controls and monitored the changing views in preparation for an imminent landing.

As the selected landing area came into view the Captain emotionally shouted. "Christ, that bloody bubble craft is there."

"What the hell is that doing their, asked the Prof, what is it, why is it there?"

"That we will have to find out when we land." Answered the Captain.

"Do you think it would be safe to land?"

"Well yes, but we'll stay on board for a while and see if there is any activity."

"Yes. Said James. It's definitely some kind of space craft. I'll get Quanto to try to communicate with it."

Lena showing her concern, said. "What happens if we can't land?"

"That's very doubtful. Commented the Captain. We are not aware of any danger that could interfere with our landing."

The Navigator proceeded to adjust the landing trajectory as the planets sandy rocky surface came into view, they were now minutes from landing.

The Shuttlecraft landing was easier than anticipated, it settled down slowly into the sandy surface and levelled up without too much propulsion activity.

The Captain leaned back in his seat and quietly said. " All crew and passengers sit tight, keep quiet and see if there is there is any activity or response from that bubble craft." Which was now in full view about 50 metres away.

After a few minutes James said." Quanto cannot communicate or observe any form of activity. I suggest that we allow it to leave the shuttlecraft and take a closer look at the bubble craft."

"Yes, good idea. Said the Captain. If there are any aliens, they will probably want to communicate and show themselves."

Quanto walked towards the bubble craft, stopped and stood about 3 metres in front of it. There was nothing, no movement or sound just the still silence.

On the shuttlecraft that all sat quietly watching a very unusual scene of a robot standing in front of a bubble spacecraft for what seemed like a long time.

Suddenly, Quanto was seen to raise his arm and step back as a panel in the lower part of the bubble slid open and a robot emerged and appeared two greet Quanto.

The Captain, crew and all passengers, all gasped with astonishment and all thought and said the same thing. *"Gosh, that's one of our rogue robots."*

Quanto turned, gesticulated and sent a message. *"This is the robot spacecraft from your earth, they want and need to help you to establish yourself on this planet."*

"How did they know where we were going"? Asked the Captain. His concern sounding in his voice.

"They picked up a message from your Spaytech observation team they sent you a message to change your trajectory to avoid the power and attraction of the black hole magnetic field which you were fast approaching."

"They obviously want to help us in some way, can you tell them that we are about to disembark."

As the Captain and the Prof approached the robot spacecraft a group of three robots emerged and stood there waiting for them to approach.

Quanto's metallic voice again shattered the pregnant silence. *Both the Captain and the Prof are concerned and curious about your presence here. They need to know your intentions."*

The leading robot faced the Prof and Captain and said. *"We have come here to support and work with you and hopefully establish an environment to physically support your human needs."*

"Well thank you for that. But how did you get here in that, it looks just like a semitransparent bubble? Said the Captain.

"You have a lot to learn, this spacecraft is a third of the weight of your enormous flying saucer, and we travel three times faster than your craft with unlimited solar fuel storage, and we can manoeuvre and land on any planet we choose without any complications."

The Captain nodded his head, looked at the Prof and said. "Perhaps, they do have something to teach us."

The three robots silently appeared to communicate with each other. The leading robot, moved forward, turned and gesticulated to their spacecraft and said.

"This craft is made up completely of Perovskite solar cells and its construction is several solar panel generators. You will need to produce several spacecraft of similar construction if you intend to evacuate the earthly human population. If you don't evacuate all life will eventually be eradicated as the earth will become uninhabitable for humans and mammals unless drastic action is taken to change your living, productive and working environment. Which so far you have found impossible to achieve, so here you are on an unknown and alien planet which we anticipate could be scientifically and biologically worked to create your new home."

The Prof was more than surprised and asked. "What is Perovskite?"

"A mineral, called Perovskite, can be used to manufacture translucent and very effective solar reception panels...the robot then continued... and it will be possible for this planet to become a green planet like yours, but you need to work with us to get it started by importing your biological earth. And start a vegetation growth that will then create the oxygen needed for the human environment.

The Prof and James waited for Quanto to react to the robots' suggestions.

Quanto turned and said. *They've obviously had time to research and check, otherwise they would not have made such a profound and suggestive statement.*

The Prof said. Gesticulating with enthusiasm.

"Well, there is a trace of oxygen in the atmosphere, it could be possible to use the biological breakup of vegetation and add it to the sandy and rocky soil. And that would create a compost and we could get the normal earth vegetation growth, which of course will be absolutely necessary if we are expecting to live here. "

"Yes, said James. We could import the topsoil with each trip, and eventually get a few square miles of arable and productive earth".

"Could be quite a long and laborious process, added the Prof's. But with our friendly robots new bubble spacecraft, it could be done, hopefully quite efficiently."

"Okay, come on, let's get back to organize and evacuate our technical team, we need to join and work with our friendly quanto robots, they here to help us to establish sustainable presence on this new planet.

Docking back on the spacecraft, the waiting crew and passengers were all aware of the encounter with the visiting quantum robots. The Prof, James and Quanto compared their notes with scientific colleagues and prepared the procedure for several more shuttlecraft landings on what was now to be their new home.

Lena's baby girl, Vanessa was the first human born on the sandy soil of planet 'Eunaton'

A few years later, Vanessa sitting in her beautiful underground room with the illuminated Perovskite solar cell domed roof the light is continually charging and producing the electric power needed to continually charge and produce the whole of the electric power needed to run her environmental eco-home.

It's now 2039, and Vanessa, the first human child born 1n 2021 on their new world. 'Eureka' opens her notebook and writes the first sentence of her story from her mother's historic Journal... *After several years of cleanup and evacuation Planet earth now several million light years away can be seen and known as the most peaceful and beautiful planet in the universe.*

She pauses for a moment, touching her lips with her pen, reads and then continues to copy her mother's expressive note's...

Our old earth world is now inhabited by robotic humanoids that enjoy a life in tune with the quantum physical nature of the cosmic universe living off the power of natural light and sun in a beautiful, creative world of natural processes of a solar environment. An idyllic robotic life enhanced with the mental attitudes and faculties of the now advanced robotic semi -human brain... now free from the reliance of the binary system and completely evolved with the quantum physical vibrations of the universal starscape of our wonderful interplanetary universe.

She again paused, picked up her mother's journal and read out loud.

" Some !! would say that we had become our own universal God. "

Outside on the surface of 'Eureka' can be seen numerous Perovskite solar cell domed roofs of the eco-environmental homes, offices and factories built and maintained with the cooperation of our humanoid companions. The activity of the human efforts together with the cooperation of the now friendly robotic humanoids were successfully building and producing a mutually acceptable living environment on what was to become our new human world.

Back on earth every biological living creature had gone, those humans that remained had died and all had been buried back into the earth where they had originated from, and the earth world now devoid of the obsessive polluted use of oil, gas, and the excessive heat and flooding was now just a memory. Virtually a new world, everything was either growing, living and naturally powered by the sun, as it was meant to be. The humanoid robots had worked through the elimination of the humans, animals and controlled the catastrophic climatic changes and been able to harness the natural power of the day and sun light.

With the elimination of the living biological world of human flesh, blood, bacterial viruses and human animal digestion was no longer contaminating the world's environment.

With the natural oxygenated, climatic rainfalls, maintaining the Earth's warm, cool and cold seas together with the natural night and daytime system continuing to maintain a seasonal world for the nonbiological mechanical humanoids to progressively continue to keep a clean and climatic control of their now inherited beautiful planet Earth.

Thank you for reading 'robotic humanoids.'
Please leave a review.
Other books originally written under Steve's pseudo-name Steve Earle.
Sex Factor International. Book 1
Sex Factor International. Book 2.
Sex Factor International. Book 3.
A walk-in time.
In a philosophical mood.
Beyond acceptance.
Website
www.evansteve.com[1]
Email
evans.steve@yahoo.co.uk
You've finished. Before you go...
Tweet/share that you finished this book
Rate this book.

---

1. http://www.evansteve.com/

Don't miss out!

Visit the website below and you can sign up to receive emails whenever Steve Earle publishes a new book. There's no charge and no obligation.

https://books2read.com/r/B-A-REZR-OISAC

Connecting independent readers to independent writers.

Also by Steve Earle

Robotic Humanoids.